Computer Modeling of Matter

Computer Modeling of Matter

Peter Lykos, EDITOR
Illinois Institute of Technology

Based on a symposium sponsored by the ACS Division of Computers in Chemistry at the 175th Meeting of the American Chemical Society, Anaheim, California, March 14, 1978.

ACS SYMPOSIUM SERIES **86**

AMERICAN CHEMICAL SOCIETY
WASHINGTON, D. C. 1978

Library of Congress CIP Data

Computer modeling of matter.
(ACS symposium series: 86 ISSN 0087-6156)

Includes bibliographies and index.

1. Molecular theory—Data processing—Congresses.
2. Matter—Properties—Data processing—Congresses.
I. Lykos, Peter George. II. American Chemical Society. Division of Computers in Chemistry. III. Series: American Chemical Society. ACS symposium series; 86.

QD461.C632 542'.8 78-25828
ISBN 0-8412-0463-2 ASCMC 8 86 1–272 1978

PRINTED IN THE UNITED STATES OF AMERICA

ACS Symposium Series

Robert F. Gould, *Editor*

FOREWORD

The ACS Symposium Series was founded in 1974 to provide a medium for publishing symposia quickly in book form. The format of the Series parallels that of the continuing Advances in Chemistry Series except that in order to save time the papers are not typeset but are reproduced as they are submitted by the authors in camera-ready form. Papers are reviewed under the supervision of the Editors with the assistance of the Series Advisory Board and are selected to maintain the integrity of the symposia; however, verbatim reproductions of previously published papers are not accepted. Both reviews and reports of research are acceptable since symposia may embrace both types of presentation.

CONTENTS

PREFACE

The ever increasing power of computers and the continuing decrease in their cost enable the chemist to construct increasingly sophisticated mathematical models of bulk matter—both at equilibrium and changing in time—from a molecular perspective.

Thanks to the pioneering work of Bernie Alder and others who developed the method of molecular dynamics and to the Monte Carlo method of Metropolis used for equilibrium data, as the size and speed of computers increased it became feasible for Aneesur Rahman and Frank Stillinger (*J. Chem. Phys.* (1971) **55**, 3336) to apply the method of molecular dynamics to the most important and complex bulk matter of all—liquid water. That seminal paper sparked a great interest in modeling on the part of chemists. The important discovery that a mathematical model whereby one averages the individual properties of a few hundred interacting molecules suffices to assess the bulk properties of many important systems suggests that chemists now can build more effective bridges between atomic and molecular physics on the one hand and surface chemistry and chemical thermodynamics and kinetics on the other.

In the recent "Science Update: Physical Chemistry" in *Chemical and Engineering News* ((1978) June 5, p 20–21), the impact of the computer primarily as an aid to modeling was highlighted as the most pervasive and important factor influencing physical chemistry today. In that article Mitch Waldrop wrote, "So important have the big computers become that number crunching and theoretical chemistry often seem synonymous. The applications can be divided into three broad areas: quantum chemistry, chemical dynamics, and statistical mechanics. If computers were big enough, those three would form a logical sequence for the complete ab initio calculation of the properties of bulk matter."

The number and range of computer-based models of bulk matter has increased rapidly. An important international conference (with proceedings), "Computational Physics of Liquids and Solids" held April, 1975, at Queen's College in Oxford, involved 34 papers that displayed a broad range of chemical phenomena being studied in this manner. Just a year and a half later the Faraday Division of the Chemical Society, London, held a two-day symposium on "Newer Aspects of Molecular Relaxation Processes" at the Royal Institution, London. The so-called 'experimental technique of examining motions in a computer-generated

liquid' was considered together with experimental methods, with theoretical models of relaxation processes, and with the hydrodynamics of rotation in fluids.

Two books have appeared which, while not presenting a unified view of the field nor a critical assessment of the literature, do provide the interested scientist with an entree to the use of computer simulation of liquids ("Theory of Simple Liquids," J. P. Hansen, I. R. McDonald, Academic, 1976; "Atomic Dynamics in Liquids," N. H. March, M. P. Tosi, Halsted (Wiley), 1977). The impact on statistical mechanics as a discipline is reflected in the two-volume work, "Statistical Mechanics," B. J. Berne, Ed., Part A, Equilibrium Techniques, and Part B, Time-Dependent Processes, Plenum, 1977. Indeed, in his Nobel prize address, I. Prigogine made several references to the use of computer simulations as an aid to development of his Nobel prize winning work on the macroscopic and microscopic aspects of the second law of thermodynamics (*Science* (1978) **201,** 777–785).

Enough experience has been gained through the design and testing of such models that chemists interested in gaining insight into particular chemical systems are beginning to apply the models in a rather sophisticated manner. For example, William Jorgensen, a theoretical organic chemist interested in solvation effects in organic chemistry, has begun by looking at liquid hydrogen fluoride in a paper to be published in December, 1978, *J. Am. Chem. Soc.* Indeed theorists are organizing and presenting their theories with careful attention to how they might be applied using a computer (for example, *see* "Simulation of polymer dynamics 1. General theory and 2. Relaxation rates and dynamic viscosity," M. Fixman, *J. Chem. Phys.* (1978) **69,** p 1527, 1538).

The call for contributions to this symposium has resulted in an interesting collection of eighteen chapters which provide the reader with a variety of touchstones ranging from the first chapter, a straight-forward application by K. Heinzinger of the Rahman–Stillinger method to an aqueous solution of sodium chloride, to the last chapter, an overview from microphysics to macrochemistry via discrete simulations including chemical kinetics.

The other chapters include examination of an algorithm assessing the importance of three-body interactions and of algorithms reducing machine requirements with regard to size of store and speed.

Most of the work done to date has been based on classical mechanics. The chapter by M. H. Kalos brings home the fact that one needs to examine the validity of such models from within the framework of quantum mechanics. The interesting phenomenon of collective modes of motion is exemplified by M. L. Klein's chapter on collective modes in solids. The liquid–vapor interface is addressed in two chapters, one by Rao and Berne and the other by Thompson and Gubbins.

Sundheim's chapter on high field conductivity emphasizes that the modeler is free to impose whatever simulated physical conditions he pleases and, in fact, can achieve in his computer experiment conditions which would be extremely difficult, if not impossible, to achieve in the laboratory.

Finally, the Chester, Gann, Gallagher, and Grimison chapter demonstrates that the computer mystique is being displaced by a hard-headed appraisal by chemists of what computer enhancements are possible with existing technology. Through the addition of a peripheral device (designed to do floating-point arithmetic very rapidly) to the campus large-scale data-processing machine, it was possible to improve the cost effectiveness of the total system about eightyfold when used for computer simulation of matter.

The growing sophistication of matter modeling as a result of advances in computer size, speed, and cost effectiveness may well have as its greatest impact, however, the enhancement of communication among those interested in any particular physical system.

Illinois Institute of Technology
Chicago, Illinois
October 13, 1978

PETER LYKOS

1

Molecular Dynamics Simulations of Liquids with Ionic Interactions

K. HEINZINGER, W. O. RIEDE, L. SCHAEFER, and GY. I. SZÁSZ

Max-Planck-Institut fuer Chemie (Otto Hahn-Institut), Mainz, Germany

Molecular dynamics (MD) simulations of liquids with ionic interaction have so far been performed for molten salts and aqueous electrolyte solutions. The characteristic problem for this kind of simulation are the far ranging Coulombic forces.

The first preliminary MD calculations for molten salts have been reported by Woodcock in 1971 (1). The large amount of work published in the meantime has been reviewed by Sangster and Dixon (2). In the case of aqueous electrolyte solutions only preliminary results for various alkali halide solutions have been published so far (3).

The more advanced state of the art for the molten salts leads to a concentration of the effort on the improvement of the algorithm for the integration of the equation of motion necessary to calculate dynamical properties with still higher accuracy. One example where very high accuracy is needed is the calculation of the isotope effect on the diffusion coefficients. In the section on molten salts below one way to improve the algorithm is discussed and the progress is checked on the mass dependence of the diffusion coefficients in a KCl melt.

Results with a certain degree of reliability from MD simulations of aqueous solutions reported up to now are restricted to structural properties of such solutions. In the section on aqueous solutions below very preliminary velocity autocorrelation functions are calculated from an improved simulation of a 0.55 molal NaCl solution. The problem connected with the stability of the system and the different cut-off parameters for ion-ion, ion-water and water-water interactions are discussed. Necessary steps in order to achieve quantitative results for various dynamical properties of aqueous electrolyte solutions are considered.

Molten Salts

The Hamilton differential equation system can be solved

0-8412-0463-2/78/47-086-001$07.00/0

numerically for a N particle system with a simple "leap-frog" algorithm as used by Verlet (4) or by a predictor-corrector algorithm. The most important criterion for the choice of the algorithm is the numerical stability. The final decision follows from necessary accuracy with which for example the energy is preserved, which in turn is determined by the desired accuracy of the transport properties, say self diffusion coefficient. In the simulated system of molten KCl it has been investigated how the total energy, the total momentum, and the velocity autocorrelation functions varies with different algorithms.

The MD calculation for the KCl-system was carried out at the melting point (1043K). The basic cube contains N=2·108 particles and the cube length S is than calculated to be 20.6A. For the pair potential the Born-Mayer-Huggins potential

$$\varphi(r) = \pm er^{-1} + b\exp(-Br) + Cr^{-6} + Dr^{-8} \qquad (1)$$

is used with the parameters given by Tosi and Fumi (5), listed in Table I.

Table I

Parameters for the Born-Mayer-Huggins Potential

	b	B	C	D
	10^{-12}erg	A^{-1}	10^{-12}erg A^6	10^{-12}erg A^8
$K^+ - K^+$	1991.67768	2.967	- 24.3	- 24.0
$K^+ - Cl^-$	1224.32459	2.967	- 48.0	- 73.0
$Cl^- - Cl^-$	4107.95500	2.967	- 145.5	- 250.0

Three different algorithms were investigated. In the first version (I) the Coulombic energies and forces were evaluated with the use of the erfc part of the Ewald method (6) only. In the other two versions (II and III) the full Ewald method was employed. In all versions the separation parameter η, the number of nearest neighbours n_η, and the maximum value of the reciprocal lattice vectors $h1^\eta$(not occuring in version I) were chosen to be:

$$\eta = 0.175, \quad n_\eta = 6, \quad h_1 = 8.$$

The algorithm which was used in version I is a so called "leap-frog" algorithm:

$$\underline{r}_i(n+1) = \underline{r}_i(n-1) + 2\Delta t\, \underline{v}_i(n-1) + (\Delta t^2/m_i)\, \underline{F}_i(n) \qquad (2)$$

$$\underline{v}_i(n) = \underline{v}_i(n-1) + \left[\underline{v}_i(n-1) - \underline{v}_i(n-2)\right] + (\Delta t^2/m_i)\left[\underline{F}_i(n) - \underline{F}_i(n-1)\right] \quad (3)$$

Here $\underline{r}_i(n)$, $\underline{v}_i(n)$, and $\underline{F}_i(n)$ are the position, velocity and force of particle i at time $t = n\Delta t$ respectively and m_i are the masses of the ions. The time step length was chosen to be $0.5 \cdot 10^{-14}$ sec. The behavior of the total energy for version I is shown in Figure 1 where $(\Delta E/E) = 6 \cdot 10^{-3}$.

A change from the "leap-frog" algorithm to a predictor-corrector algorithm has been made in version II. For the position vectors one can write:

$$\underline{r}_i^p(n+1) = \underline{r}_i^c(n-1) + 2\,\Delta t\, \underline{v}_i(n-1) + (\Delta t^2/m_i)\left[\underline{F}_i(n) + \underline{F}_i(n-1)\right] \quad (4)$$

$$\underline{r}_i^c(n+1) = \underline{r}_i^c(n-1) + 2\,\Delta t\, \underline{v}_i(n-1) + (\Delta t^2/4m_i)\left[\underline{F}_i(n+1) + 4\underline{F}_i(n) + 3\underline{F}_i(n-1)\right] \quad (5)$$

Here the indices p and c indicate predicted and corrected values respectively. The formula for the velocities has been changed slightly compared to the formula (3):

$$\underline{v}_i(n) = \underline{v}_i(n-1) + (\Delta t/2m_i)\left[\underline{F}_i(n) + \underline{F}_i(n-1)\right] \quad (6)$$

With this version II, two simulations with different time steps were carried out starting from the same configuration ($0.5\ 10^{-14}$s and $0.25\ 10^{-14}$s) over a time interval of 1.2 ps. The total energies for both runs are shown in Figure 1 . The value $(\Delta E/E)_{0.5}$ has decreased compared to the one of version I by one order of magnitude. Moreover $\Delta E/E$ decreased further by a factor 0.5 when the time step was shortened to $0.25 \cdot 10^{-14}$s.

In version III the positions and velocities have been treated with the same predictor-corrector algorithm. This means that $\dot{\underline{F}}_i(t)$, the derivative of the forces with respect to time, appears in the formula for the velocities.

$$\underline{r}_i^p(n+1) = \underline{r}_i^c(n-1) + 2\,\Delta t\, \underline{v}_i^c(n-1) + (2\Delta t^2/3m_i)\left[2\underline{F}_i(n) + \underline{F}_i(n-1)\right] \quad (7)$$

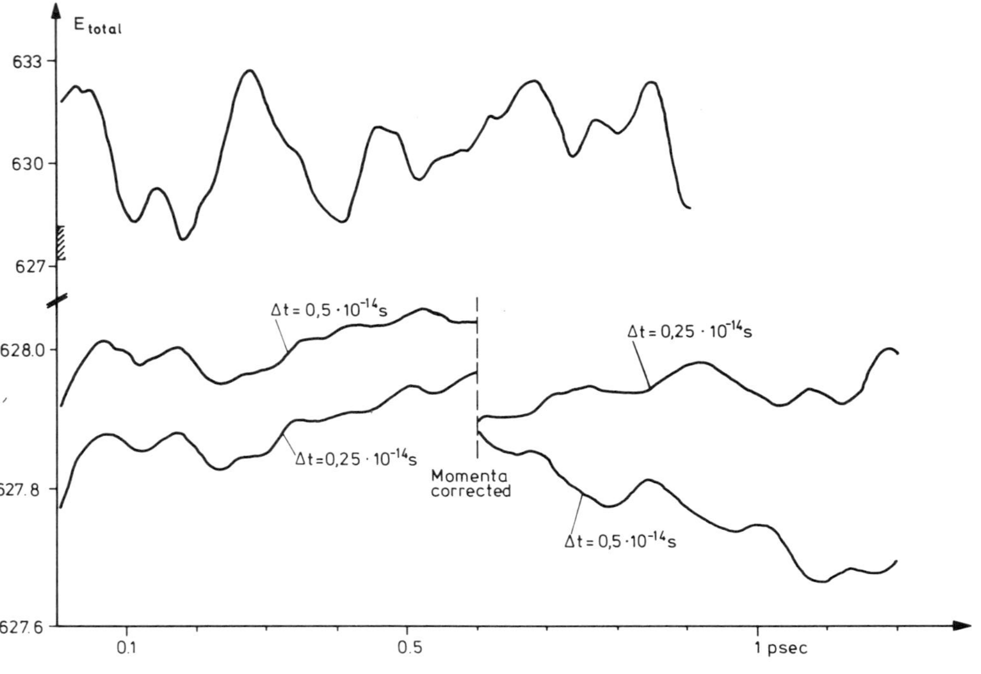

Figure 1. Total energy of version I (upper curve) *in units of* 10^3 *J/mol and of version II* (lower curve) *with a energy scale enlarged by a factor of 4¾. The length marked by (////////) in the upper scale corresponds to the interval of the lower scale. The time scale starts after an equilibration run over 7.2 psec from a roughly equilibrated starting configuration.*

$$\underline{r}_i^c(n+1) = \underline{r}_i^c(n-1) + 2\,\Delta t\,\underline{v}_i^c(n-1) + (\Delta t^2/6m_i)\left[\underline{F}_i(n+1) + 6\underline{F}_i(n) + 5\underline{F}_i(n-1)\right] + (\Delta t^3/24m_i)\left[3\dot{\underline{F}}_i(n-1) - 2\dot{\underline{F}}_i(n) - \dot{\underline{F}}_i(n+1)\right] \quad (8)$$

and for the velocities:

$$\underline{v}_i^p(n) = \underline{v}_i^c(n-2) + 2\,\Delta t/m_i\,\underline{F}_i(n-2) + (2\Delta t^2/3m_i)\left[2\dot{\underline{F}}_i(n-1) + \dot{\underline{F}}_i(n-2)\right] \quad (9)$$

$$\underline{v}_i^c(n) = \underline{v}_i^c(n-1) + (\Delta t/2m_i)\left[\underline{F}_i(n) + \underline{F}_i(n-1)\right] + (\Delta t^2/12m_i)\left[\dot{\underline{F}}_i(n-1) - \dot{\underline{F}}_i(n)\right] \quad (10)$$

It has to be emphasized that there is a basic difference between the algorithms of version II and III. In version II the velocities were not handled in the same way as the positions by a predictor-corrector algorithm. This is less time consuming per time step and might be sufficient for a given accuracy in the total energy for example.

The values of $(\Delta E/E)_{0.5}$ of Version III decreased by two orders of magnitude compared to version I. When the time step was shortened by half, $(\Delta E/E)_{0.25}$ decreased one order of magnitude more. Finally $\Delta E/E$ decreased by three orders of magnitudes compared with $\Delta E/E$ of version I. An compilation of all $\Delta E/E$ is given in Tab.II according to the different versions and time steps.

Table II

Values of $\Delta E/E$ according to the different versions

version	I	II		III	
$\Delta t \cdot 10^{-14}$s	0.5	0.5	0.25	0.5	0.25
$\Delta E/E$	$6\ 10^{-3}$	$6\ 10^{-4}$	$3\ 10^{-4}$	$3\ 10^{-5}$	$2\ 10^{-6}$

The fluctuation of the kinetic energy is about ± 2500 J/mol corresponding to a fluctuation of temperature of about $\pm$ 100K. Compared with the total energy it follows $(\Delta E_{kin}/E)_{0.25} = 4 \cdot 10^{-3}$. This shows that the fluctuations of the kinetic energy are three orders of magnitude larger than the fluctuations of the total energy and hence the fluctuations of the kinetic energy are almost compensated by the corresponding fluctuations of the potential energy.

In order to keep the computing time small for each time step, care has been taken to calculate the right hand side (r.h.s) of the differential equation system only once per time step. This means that the predicted values $\underline{r}_i^p$ should not be too different from the corrected values $\underline{r}_i^c$. With version III

$$|\underline{r}_i^p - \underline{r}_i^c| < \varepsilon \tag{11}$$

for each component (x,y,z), each time step, and all particles i where $\varepsilon = 10^{-6}$A. Moreover for about 70% of the particles $\varepsilon = 10^{-7}$A could have been chosen. From eq. (7) and (8) follows

$$\Big| \underline{F}_i(n) - \underline{F}_i(n+1) - \left[\underline{F}_i(n+1) - \underline{F}_i(n) \right] - (\Delta t/4)\left[3\dot{\underline{F}}_i(n-1) - 2\dot{\underline{F}}(n) - \dot{\underline{F}}_i(n+1) \right] \Big| < 6\varepsilon m_i/\Delta t^2 \tag{12}$$

This inequality can be estimated if one replaces the differentiation by the corresponding ratio of the differences which yields:

$$|\Delta \underline{F}_i(n) - \Delta \underline{F}_i(n-1)| < 4\varepsilon\, m_i/\Delta t^2 \tag{13}$$

with $\underline{F}_i(n) - \underline{F}_i(n-1) = \Delta \underline{F}_i(n)$. Eq.(13) shows that the change of the force change within the given time step determines the ratio $\varepsilon/\Delta t^2$ for constant mass.

The desired accuracy of the particle positions requires an adequate time step Δt and determines the accuracy of the forces. Eq.(13) would be meaningless if the errors of the forces were as large as the changes of the force change, because the correction added to the predicted value of $\underline{r}_i$ would be of the order of the errors of the forces. However, ε has to be small enough because one wants to calculate the forces with the predicted values of $\underline{r}_i$ only.

An analogous equation to eq.(11) is given for the velocities with an ε' which is different from ε.

$$|\underline{v}_i^p - \underline{v}_i^c| < \varepsilon' \tag{14}$$

For ε' a value of 20 cm/s has been found which corresponds to a maximum change in the positions of $5 \cdot 10^{-6}$A.

This shows that the positions of the particles will be slightly changed by taking the corrected velocities rather than the predicted values in eq.(8). Moreover ε' corresponds to a maximal temperature difference ΔT which can be estimated by

$$\Delta T < (\bar{m}/3k)\varepsilon' \left[\varepsilon' + 2\, (3kT/\bar{m})^{1/2} \right] \tag{15}$$

Here $\bar{m}$ is the average mass, k Boltzmann's constant, and T the temperature at the melting point. The value of the r.h.s. of eq. (15) is 0.5 K which is 5 ‰ of the temperature fluctuations of about $\pm$ 100 K. This means that the temperature fluctuations are mainly determined by the change of the velocities in time rather than by the corrections of the velocities.

The absolute value of the total momentum $|\underline{P}|$ of the system is shown in Figure 2. It decreased by almost one order of magnitude between version I and II. Almost no change occured in $|\underline{P}|$ comparing version II with III. Therefore only $|\underline{P}|$ of version III has been plotted in Figure 2. However, a decrease by a factor of 4 showed up when the time step was reduced. Additionally, $|\underline{P}|$ was reduced by a factor of about 3 when the initial configuration also included the differentiation of the forces. This total momentum curve is marked with a capital A in Figure 2. A more precise analysis of the total momentum make it reasonable to assume

$$\underline{P} = \int_0^t \underline{F}(t')\, dt'. \qquad (16)$$

Hence the total force $\underline{F}(t)$ does not vanish completely for each time step, the total momentum is the sum of the deviations of the total force from zero, rather than the total momentum of the system in a physical sense. $|\underline{P}|$ was brought to zero for the equilibrated initial configuration. The extrema of the total momentum components in Figure 3 correspond to the zeros of the total force components at the same time points. Moreover the forces of all particles have been separated into the different parts of the pair potentials, eq.(1), and have been summed up separately,

$$\underline{F} = \sum_i^N \underline{F}_i = \sum_i^N \underline{F}_i^c + \sum_i^N \underline{F}_i^{ex} + \sum_i^N \underline{F}_i^{(6)} + \sum_i^N \underline{F}_i^{(8)}. \qquad (17)$$

Here the indices c, ex, 6, and 8 indicate to which part of the pair potential the forces belong. The calculation has shown that all terms of eq.(17) except the coulomb term are smaller than $3 \cdot 10^4$ dyn/mol for each component while the coulomb term is seven orders of magnitude larger, $3 \cdot 10^{11}$ dyn/mol. Consequently, the deviation of the total force from zero is determined by the sum of coulombic forces.

The velocity autocorrelation functions and from them the self diffusion coefficients have been calculated from all versions for the normal masses as well as for the mass of the anion reduced by 26%. They are shown together in Figure 4. The velocity autocorrelation functions in version I do not behave qualitatively correct. If the anion mass decreases, the first zero of the anion correlation function should shift to smaller time, while the zero of the unchanged cation correlation function should slightly shift to larger time. The qualitatively correct behaviour of the velocity autocorrelation functions results from version II and III.

The behaviour of the total energy, kinetic energy, the instantaneous value of $P^+(t) = (2E_{kin} - \psi)/3V_m$, and the velocity autocorrelation functions of the system have been investigated with a change of the cut-off radius in the interaction which belongs to the pair potential part $1/r^6$. Here is the virial of the system calculated by

$$\psi = - \sum_{i,j}^N r_{ij}\, F(r_{ij}) \qquad (18)$$

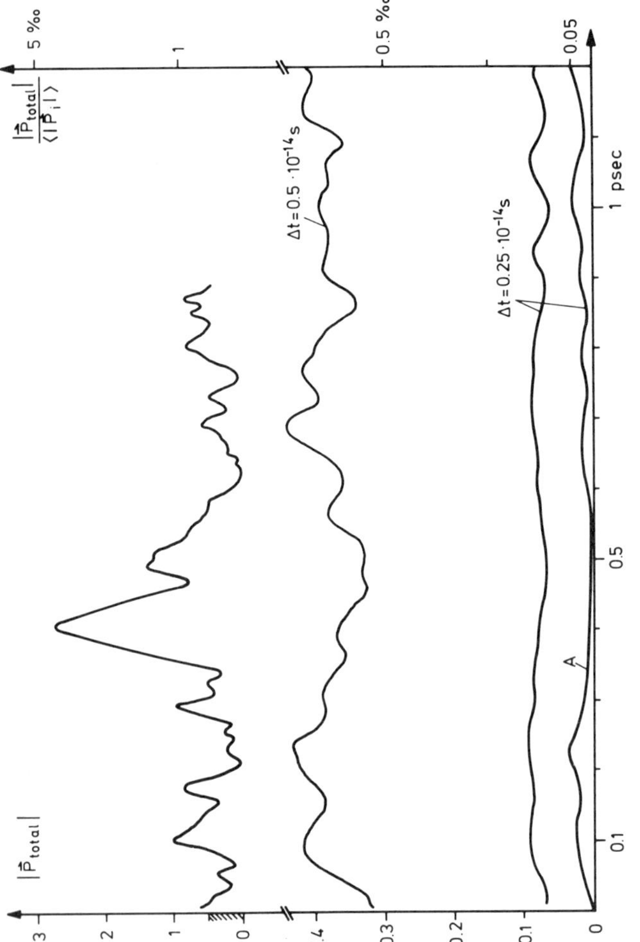

Figure 2. Absolute total momentum in units of 10^4 cm g/sec mol on the left-hand scale, while the right-hand side relatively gives the absolute momentum to the averaged momentum per particle. The upper curve for version I, the lower curves for version III, with a scale enlarged by a factor of $4^3/_4$.

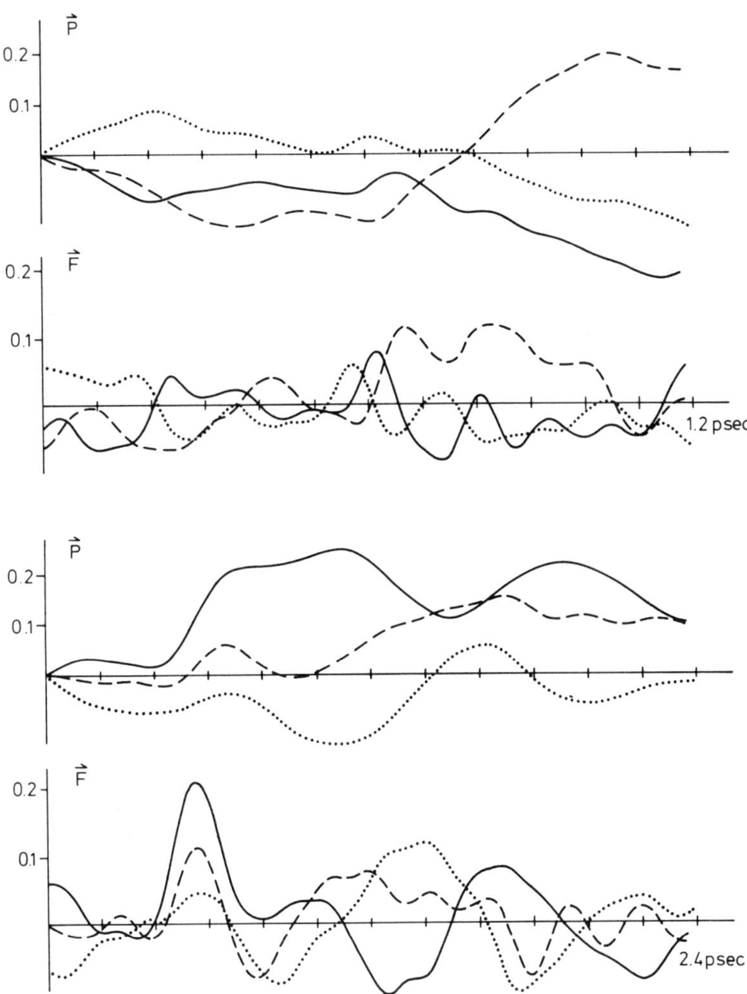

Figure 3. Total momentum P *in units of* 10^4 *cm g/sec mol, total force* F *in units of* 10^{12} *dyn/mol for the* x,y,z *components (solid, dashed, and dotted line, respectively)*

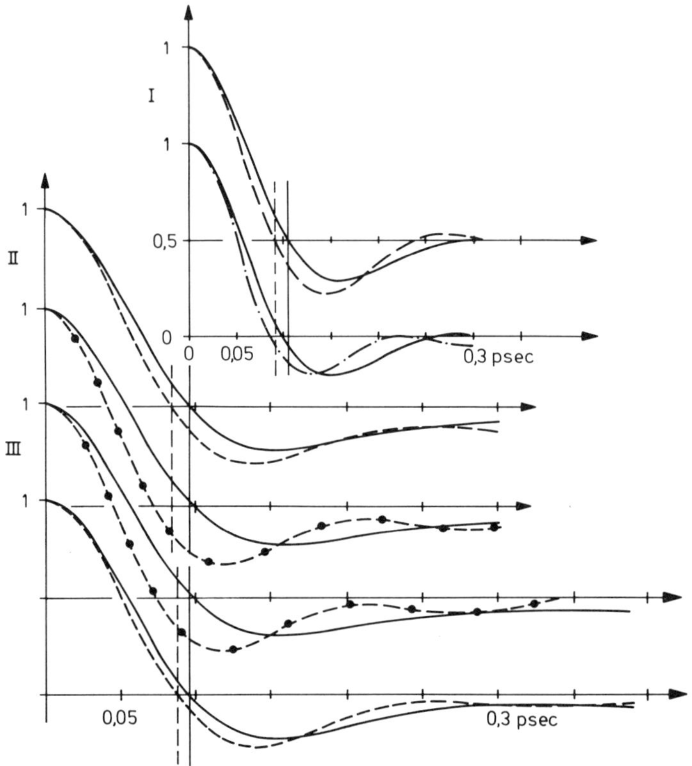

Figure 4. Velocity autocorrelation functions from version I, II, and III for $^{39}K^{+}$ (——), $^{35}Cl^{-}$ (– – –), ^{31}Cl (·–·–), $^{27}Cl^{-}$ (-•— — —•-). Vertical lines allow zeros of the correlation functions to be compared.

No cut-off radii have been changed for the other interactions. In calculation (I) all 215 particles in the environment cube (the cube with edge of length S and particle i at its center) of particle i have been taken into account for the calculation of the force acting on particle i and the energy. In a second run (II), which was started from the same configuration, all the particles have been taken into account within the sphere which circumscribes the environment cube of particle i. The total energy decreased as expected by about 0.02%) and stayed constant within $(\Delta E/E)_{0.25} = 1.6 \cdot 10^{-6}$ over the whole run of 1440 time steps. As no iteration happened in either of the calculations I and II, eq.(13) still holds especially for the beginning of the runs. This result shows that the deviations have been of the order of the allowed uncertainties of the forces.

The kinetic energy and $P^{+}(t)$ differ more and more from each other as a function of time as is shown in Figure 5 . However, no qualitative changes of the kinetic energy nor of the quantity $P^{+}(t)$ have occured. Although the changes which have taken place in the forces $F_i^{(6)}$ are very small compared to the forces F_i themselves, a significant change appeared in the kinetic energy. Thus a small uncertainty in the forces at each time step which is not purely random or is not cancelled by summation, might produce a trend for example in the kinetic energy. Furthermore a slightly different starting configuration causes a quantitatively but not qualitatively different kinetic energy. This fact allows the choise of the starting velocity vector (3N-dimensional) independently of the time sequence of calculation of these vectors with regard to the averaging procedure for the autocorrelation functions. This is discussed in ref.(7) in more detail. However, the cut-off radius does influence the velocity autocorrelation functions as shown in Figure 6 and the corresponding time integrations which lead to the diffusion coefficient as shown in Figure 7 . The differences between the velocity autocorrelation functions of runs I and II are larger than the root-mean-square error of the mean for part C (see curve I C and II C in Figure 6 . It is obvious that this error could lead to an underestimation of the errors for the velocity autocorrelation functions, as is discussed in more detail in ref. (8). Moreover, when comparing the velocity autocorrelation functions from Figure 6 with the ones from Figure 4 for the lighter anion mass, it seems to be necessary to have a higher accuracy for larger times (say round about 0.5 psec), at least as high as in version III, in order to get a significant difference in the self diffusion coefficients.

In order to reduce these errors, it seems to be better to increase the number of particles per cube rather than to increase the number of time steps. A larger number of particles brings about another improvement besides a better statistics. The necessity of introducing periodic boundary conditions (discussed in detail in many places, e.g.(1,2,8,12)) has some influence on the

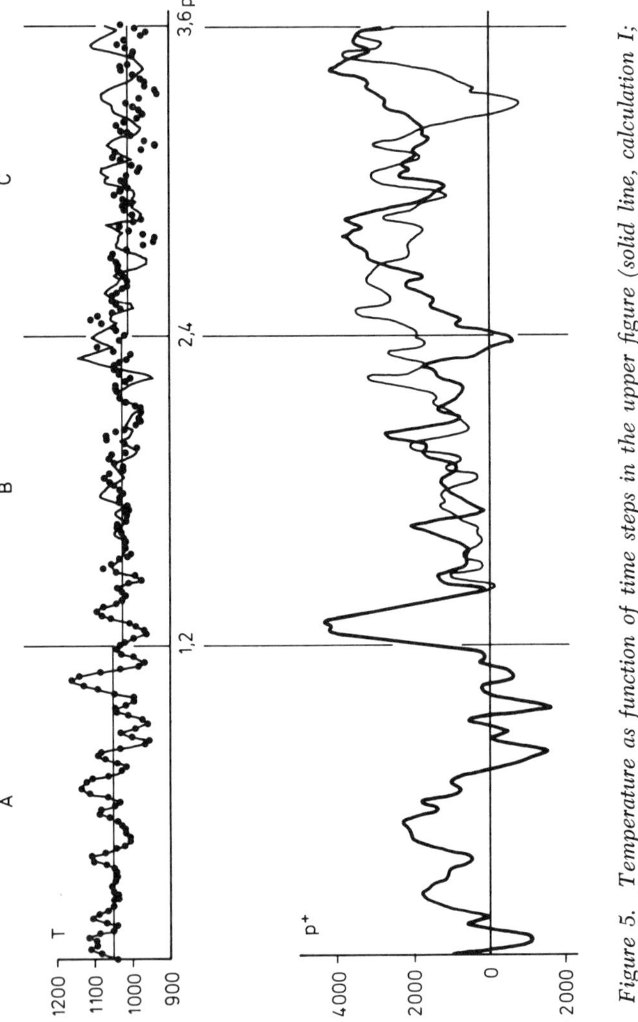

Figure 5. Temperature as function of time steps in the upper figure (solid line, calculation I; dots, calculation II). P+ *as function of time steps in the lower figure (heavy solid line, calculation I; light solid line, calculation II).*

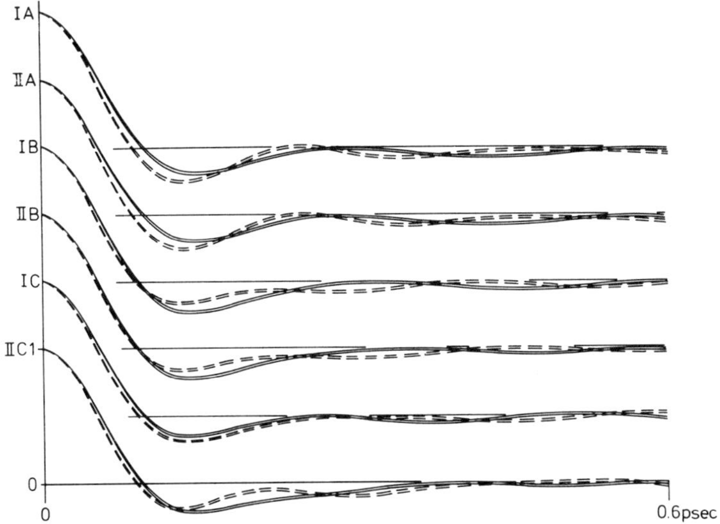

Figure 6. Velocity autocorrelation functions plus and minus the root-mean-square error of the mean averaged over 60 starting velocity vectors for each part A, B, and C (Figure 5) of calculations I and II respectively (solid line, $^{39}K^+$, dashed line, $^{35}Cl^-$)

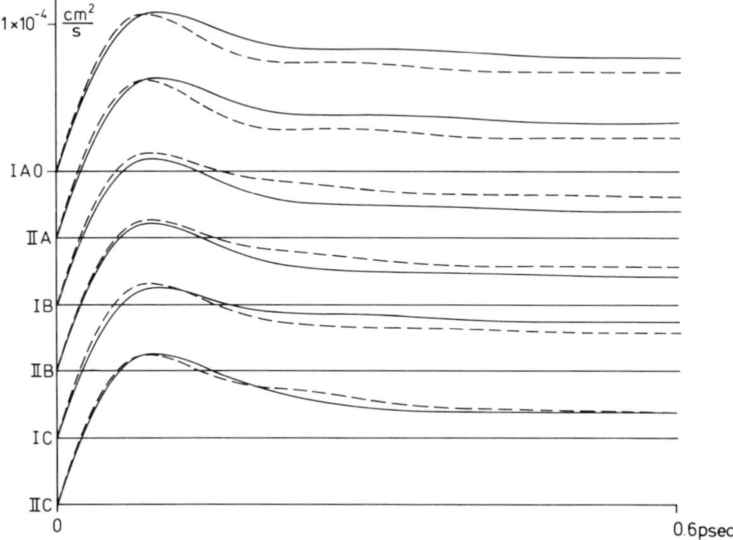

Figure 7. Integration of the velocity autocorrelation functions (Figure 6) leading to the self-diffusion coefficient for large times (approximately 0.5 psec). The nomenclature is equivalent to that in Figure 6.

system too, as can be seen for example from the different values of the virial obtained, depending whether one calculates

$$\psi = -\sum_{i,j}^{N} r_{ij}\, F(r_{ij}) \quad \text{or}$$

$$\psi = -\sum_{i,j}^{N} \underline{r}_i \cdot \underline{F}_i$$

The difference will decrease for the non Coulombic interactions if the number of particles is increased, since the number of particles whose interaction sphere contains periodic images will become smaller than the number of particles whose interaction sphere contains no periodic images. This means that the cut-off radius for the non Coulombic interaction has to be sufficiently large, so that the absolute value of the particle - furthest particle force is small enough compared to the absolute value of the force acting on a particle; but the cut-off radius also has to be small compared to the cube length S.

In the above discussion the Coulombic interaction was left out. With the long range Coulombic forces, it is possible that the periodic boundary conditions being used here will lead to behaviour not present in a real physical system, when a finite size cube is used. However, it might be possible to assess the effect of the boundary conditions on the model by increasing the number of particles per cube. At present no practical alternative more physical model has been demonstrated. (A model which may eliminate the periodicity problems has been proposed by Friedman (20). To the best knowledge of the authors this proposal has not been incorporated yet into a computer algorithm).

Aqueous Solutions

The rigid ST2 point charge model used for the simulation of pure water by Stillinger and Rahman (9) with its rotational degrees of freedom is employed for the MD calculations of aqueous solutions. It requires a different kind of algorithm than the one used for molten salts. A stabilized predictor - corrector formalism is employed which has been developed by Gear (10) for the solution of differential equations of all orders. This algorithm is based on a method by Adam and improved for an easy change of the time step length by Nordsieck (11). Pure water has been simulated with the same algorithm (12).

The method is based on the knowledge of the first N time derivatives of the variables $y_i(t)$ at a given time. For the intergration of the equation of motion

$$G_i = \left[\, y_i^{(p_i)} - f_i(y_1, y_2, \ldots, y_s) \right] \frac{(\Delta t)^{p_i}}{p_i!} \tag{22}$$

the predicted values are calculated through a Taylor series expansion. $p_i = 2$ for the centre of mass coordinates x_i and $p_i = 1$ for the Euler angles (or quaternion parameters) α_i and angular velocities. If

$$a_j = \frac{(\Delta t)^j}{j!} \quad \frac{d^j y}{dt^j} \tag{23}$$

and $\underline{a}$ are (N+1)-dimensional vectors with the components $a_o, a_1, \ldots, a_N$, then

$$\underline{a}^{(o)}_{t+\Delta t} = A\, \underline{a}_t \tag{24}$$

is the predicted value, where A is the Pascal Triangle $A_{kl} = \binom{l}{k}$. The predictor - corrector formulae have the form

$$\underline{a}^{(m+1)}_{t+\Delta t} = a^{(m)}_{t+\Delta t} + \underline{\lambda} G\, (\underline{a}^{(m)}_{t+\Delta t}) \tag{25}$$

$\underline{\lambda}^T = (\lambda_1, \lambda_2, \ldots, \lambda_{N+1})$ is chosen in such a way that the integration remains stable nummerically after n time steps. $\underline{\lambda}$ depends on the order of the integration procedure. A fifth order was chosen for the translational and a seventh order for the rotational motion. The higher order for the rotational motion is necessary because of the faster change of the Euler angles compared with the translational movement, a consequence of the small moments of inertia of the water molecule. It follows from the order of the integration procedure that the local truncation error for the centre of mass coordinates is a function of Δt^6 and for the Euler angles and angular velocities a one of Δt^8.

The situation for the simulation of aqueous solutions is demonstrated with a 10,000 time step run of a 0.55 molal NaCl solution. The basic periodic box contained 200 water molecules, 2 sodium and 2 chloride ions. The ST2 point charge model is employed and the alkali and halide ions are treated as point charges residing at the centre of Lennard - Jones spheres.

Each of the six effective pair potentials consists of a LJ term

$$V^{LJ}_{ij}(r) = 4\, \varepsilon_{ij} \left\{ (\sigma_{ij}/r)^{12} - (\sigma_{ij}/r)^6 \right\} \tag{26}$$

where i and j refer either to ions or water molecules and a Coulomb term, different for water-water, ion-water and ion-ion interactions, as given by

$$V^C_{ww}(r, d_{11}, d_{12}, \ldots) = S_{ww}(r) \cdot q^2 \cdot \sum_{\alpha,\beta=1}^{4} (-1)^{\alpha+\beta} / d_{\alpha\beta}$$

$$\underset{(-w)}{V_{+w}}(\underset{(-1)}{d_{+1}}, \underset{(-2)}{d_{+2}}, \ldots) = \underset{(+)}{-} \sum_{\alpha=1}^{4} (-1)^{\alpha} \cdot q \cdot e / \underset{(-\alpha)}{d_{+\alpha}}$$

$$\underset{\substack{-- \\ (+-)}}{V_{++}}(r) = \underset{(-)}{+}\, e^2 / r \tag{27}$$

The switching function, $S_{WW}(r)$ has been introduced to reduce unrealistic Coulomb forces between very close water molecules (12). e is the elementary charge and q = 0.23 e is the charge in the ST2 water model, d and r denote distances between point charges and LJ centres, respectively. The choice of α and β, odd for positive and even for negative charge yields the correct sign.

The LJ parameters for water-water interaction are taken from the ST2 model, and for cation-cation from the isoelectronic noble gases (13). The anion-anion parameters are derived from the cation ones on the basis of the Pauling radii as shown in detail in a previous paper (14). The LJ parameters between different particles are calculated by application of Kong's combination rules (15). All ε and σ are given in Table III.

Table III

Lennard - Jones parameters for the various pair potentials in a NaCl solution is given in A and in units of 10^{-16} erg.

	H_2O	Na^+	Cl^-
	3.10	2.92	4.02
H_2O	52.61	54.84	30.83
	-	2.73	3.87
Na^+	-	5937	28.32
	-	-	4.86
Cl^-	-	-	27.87

The classical equations of motion are integrated in time steps of $\Delta t = 1.09 \cdot 10^{-16}$ s(22).In order to take account of the different strengths of the various interactions and to keep the computer time in reasonable limits, different cut off parameters are chosen for ion-ion, ion-water and water-water interactions. The Ewald summation (6), which gives the correct Coulombic energy for an unlimited number of image ions, is used for ion-ion interactions. As the ions residing in the first and second neighbour image boxes are included in the direct calculation of the so called error function part, a suitable choice of the separation parameter allows to neglect the Fourier part of the Ewald sum. The ion-water interaction is cut off at 9.1 A, half the sidelength of the basic periodic box. The water-water interaction is cut off at 7.1 A which means that only first and second nearest neighbour molecules are included in the calculation.

The structural properties of the NaCl solution calculated from this simulation are not significantly different from preceding ones. The characteristic distances an heights of the ion-water and water-water radial pair correlation function remain almost unchanged whether a temperature control mechanism is built in or not. The radial pair correlation functions from the simulation of a 2.2 molal NaCl solution (200 water molecules and 8 ions of each kind) are compared in detail with X-ray investigations in a previous paper (14). The problem connected with the determination of hydration numbers have also been discussed previously (16). In addition it has been concluded from the MD simulations of alkali halide solutions that in the first hydration shells a lone pair orbital of the water molecules is directed towards the cation and a linear hydrogen bond is formed with the anion (17). This result is also in agreement with conclusions from experimental and other theoretical investigations.

The real aim of MD simulations is, of course, the calculation of the dynamical properties of the liquid. The velocity autocorrelation function have been calculated from this 10.000 time step run. In all cases the functions have been averaged over 360 starting vectors. The statistical quality is much better, of course, in the case of the water molecules where the ensemble average extends over 200 particles compared with 2 in the case of the ions. In Figure 8 the angular (solid line) and the translational (dashed line) velocity autocorrelation functions of the water molecules are shown. The correlation time for the angular velocity is as expected much shorter than for the translational velocity. In Figure 9 the velocity autocorrelation functions for the cation (solid line) and the anion (dashed line) are drawn. If these autocorrelation functions are integrated, in spite of all statistical uncertainties it is found that the corresponding diffusion coefficients are of the same order of magnitude as is known or estimated from experimental investigations. Therefore it is justified to continue the investigation along this line. It is obvious that much longer simulation times are required to make quantitative statements about the dynamical properties. But not only an increase in simulation time is necessary, also the simulation itself needs further improvement.

The system simulated corresponds to a microcanonical ensemble and consequently it should have a constant total energy after an equilibration time of suitable length. In Figure 10 the total energy (solid line) and the potential energy (dashed line) for the system are shown over the 10.000 time steps run. I can be seen that the increase of the total energy is equally distributed between kinetic and potential energy. The unavoidable cumulative errors due to the algorithm, which were discussed above in detail in the case of the molten salts, are not the main source of energy increase in the simulation of aqueous solutions. Responsible for the problems here are mainly the cut-off radii. Figure 11 shows the variations of the total energy of the basic box E during a simulation run of 270 Δt, which is essentially caused by the cut-off radii R_W and R_i in the evaluation of the potential energy $E_{pot}(t)$. Let us consider the expression:

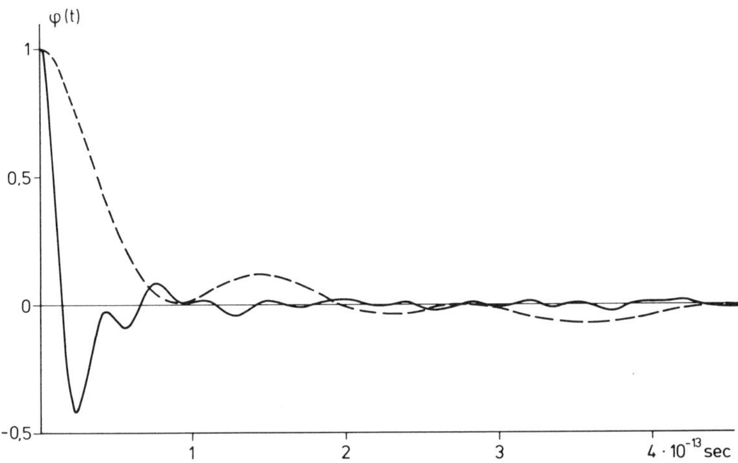

Figure 8. Normalized translational (dashed) and rotational (solid) velocity autocorrelation functions φ(t) of water in a 0.55m NaCl solution

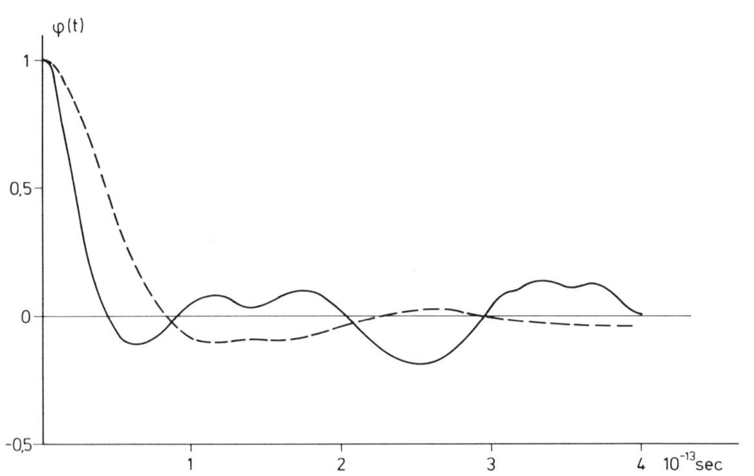

Figure 9. Normalized velocity autocorrelation functions φ(t) for cations (solid) and anions (dashed in a 0.55m NaCl solution)

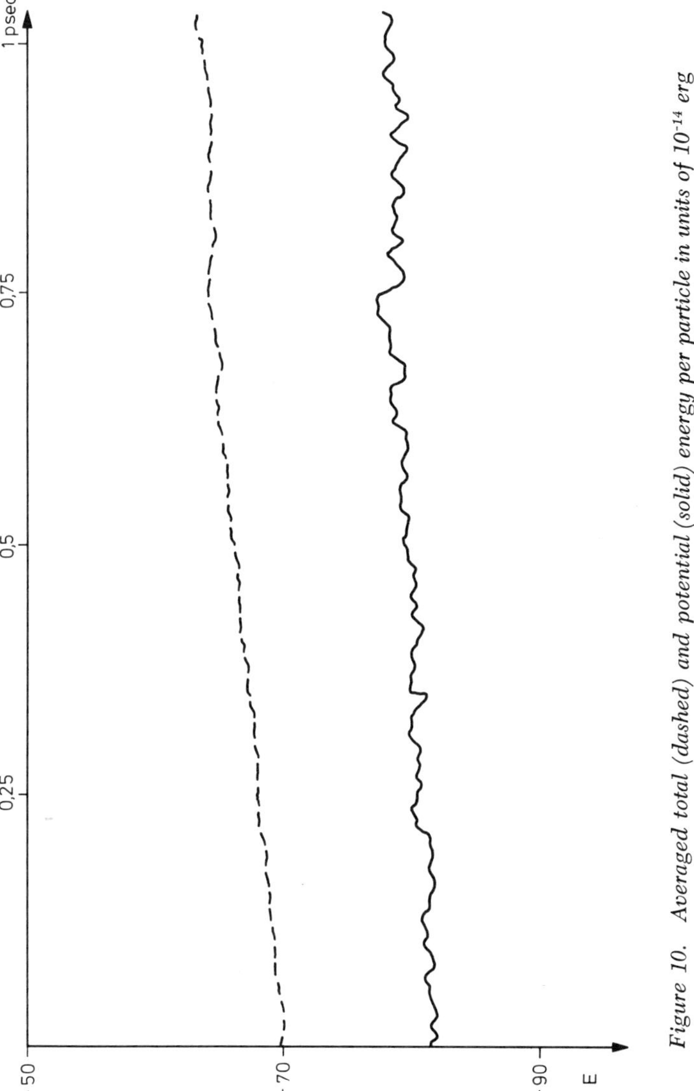

Figure 10. Averaged total (dashed) and potential (solid) energy per particle in units of 10^{-14} erg

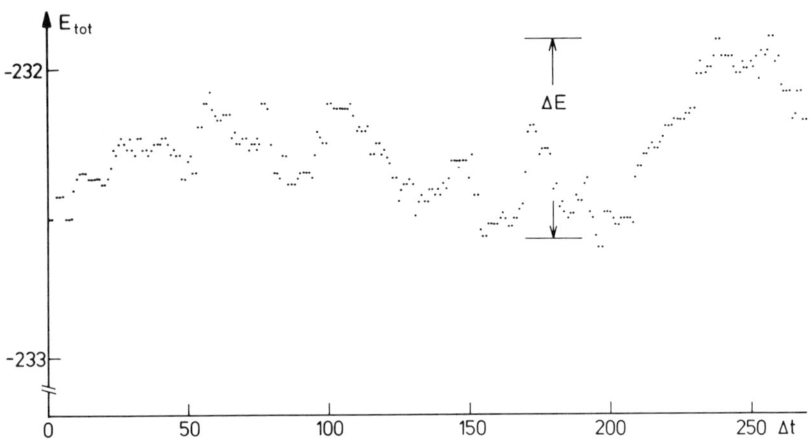

Figure 11. Variation (cut-off jumps) of the total energy of the basic box in units of 10^{-12} erg for 200 water molecules and eight NH_4Cl. The time step is $2.18 \cdot 16^{-16}$ sec and $\Delta E/E = 3 \cdot 10^{-3}$.

$$E_{pot}(t+\Delta t)-E_{pot}(t)=\sum_{i<j}^{I}\{V_{ij}(r_{ij}(t+\Delta t)-V_{ij}(r_{ij}(t))\} + \sum_{i<j}^{II} V_{ij}(r_{ij}(t+\Delta t))-\sum_{i<j}^{III} V_{ij}(r_{ij}(t)) \quad (28)$$

where the summation extends over i and j such that

$$\begin{aligned} I &: r_{ij}(t)\leq R \text{ and } r_{ij}(t+\Delta t)\leq R, \\ II &: r_{ij}(t+\Delta t)\leq R<r_{ij}(t), \\ III &: r_{ij}(t+\Delta t)> R\geq r_{ij}(t) . \end{aligned}$$

The last two terms are due to the fact that a particle i located within the spherical environment with radius R of a particle j at a certain time may leave this sphere in the subsequent time step, or vice versa. Using $|r_{ij}(t+\Delta t) - r_{ij}(t)| \ll R$ and $r_{ij}(t) \approx R$ one can estimate the last term of eq. (28) as:

$$\sum^{III} \approx \overline{V}_{ww}(R_w)\, n_w(t) + \overline{V}_{w+}(R_i)\, n_+(t) + \overline{V}_{w-}(R_i)\, n_-(t) \quad (29)$$

As the interactions depend on the orientations of the particles, averaged potentials $\overline{V}_{ww}, \overline{V}_{w+}$ and $\overline{V}_{w-}$ are considered here. $n_w(t)$ ($n_+(t)$ and $n_-(t)$) is the number of water molecules which leave one of the cut-off environments of central water molecules (ions) during a time step. Furthermore, using the number density of water, one gets

$$\begin{aligned} n_w(t) &\approx R_w^2\, (3kT/m_w)^{1/2}\, \Delta t\, (N_w - 1) = n_1 \\ n_{\pm}(t) &\approx R_i^2\, (3kT/m_{\pm})^{1/2}\, \Delta t\, (N_{\pm}) = n_2 \end{aligned} \quad (30)$$

with N_w the number of water molecules and N_i the number of ions in the basic periodic box. m_w and m_i are the masses of water molecules and ions, respectively, Table IV shows quantities from our simulations.

Table IV

Quantities relevant to energy jumps due to the cut-off radii for a 2.2 molal NH_4Cl solution.

	w-w cut-off	w-i cut-off
R	7.1 A	9.1 A
$\|V(R)\|/E$	$2 \cdot 10^{-4}$	$7 \cdot 10^{-4}$
n	1.5	$n_+ = 1/10$, $n_- = 1/14$

$1/n_2$ is the average number of time steps after which a jump occurs.

The fluctuations and the upward trend in both the translational and rotational averaged kinetic energy of the water molecules are shown in Figure 12 in Kelvin. The much faster fluctuations of the rotational energy are obvious. As expected a fast exchange of energy between rotational and translational degrees of freedom occurs. It is demonstrated in Figure 13 that the distribution of rotational energy is Maxwellian. This is also true for the other subsystems. In Figure 14 the trend in the total kinetic energy is clearly recognizable together with the much stronger fluctuations in the case of the ions where only two of each kind are in the basic periodic box. The lighter sodium fluctuates significantly faster than the chloride ion.

It is known from previous simulations at higher NaCl concentration that the upward trend in the total energy increases with concentration. As the ion-ion interaction is calculated with rather high accuracy (Ewald summation), it can be concluded that the energy increase has to be attributed to the cut off radius of the ion-water interaction.

The upward trend in energy is also a hint that during the stepwise integration of the equation of motion the stabilizing effect of the integration algorithm gets partially lost because of cumulative errors. It is shown now that inaccuracies in the force calculations may lead to a loss in the stabilizing effect of the corrector. Let us consider center of mass coordinates (com) x, and use the abreviations of eq.(28). The time step is choosen in such a way, that only one iteration is needed in the predictor-corrector formalism. Therefore one has $x_2(t) = (\Delta t)^2/2 \cdot F(t)$, and the forces F (t) are calculated with predicted com coordinates and Euler angles. Starting from an initial value $x_0(t)$, the algorithm yields for $x_0(t+\Delta t)$:

$$\begin{aligned} x_0(t+n\Delta t) &= x_0(t) + \sum_{m=0}^{n-1} \left[x_1(t+m\Delta t)+F(t+m\Delta t)\,\Delta t^2/2\right] \\ &+ \frac{\Delta t^2}{2}\left[A_n \Delta F(t+n\Delta t)+A_{n-1}\Delta F(t+(n-1)\Delta t)+ \dots \right. \\ &\left. +A_1 \Delta F(t+\Delta t)\right] + a_n\, x_3(t)+b_n\, x_4(t)+c_n\, x_5(t). \end{aligned} \tag{31}$$

The coefficients A_i and a_n, b_n, c_n can be calculated from the stabilization coefficients λ_j.

If the difference of the forces between two time steps

$$\Delta F(t+m\Delta t) = F(t+m\Delta t)-F(t+(m-1)\Delta t) \tag{32}$$

is not much larger than the inaccuracies in the evaluation of the forces. F^{err}, a stabilizing effect cannot be expected from the A_i, a_n, b_n, c_n.

For a given system, $F(t+m\Delta t)$ depends on the magnitude of the time step and can therefore easily be changed. F^{err} then

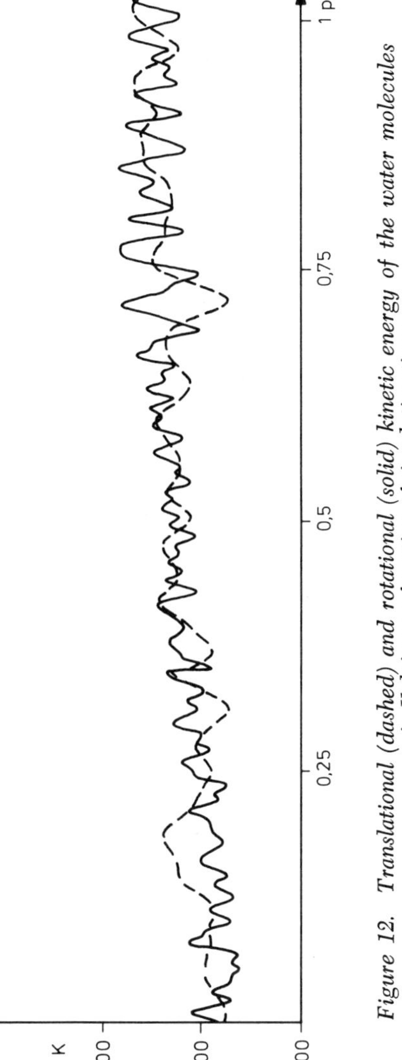

Figure 12. Translational (dashed) and rotational (solid) kinetic energy of the water molecules in Kelvin as a function of simulation time

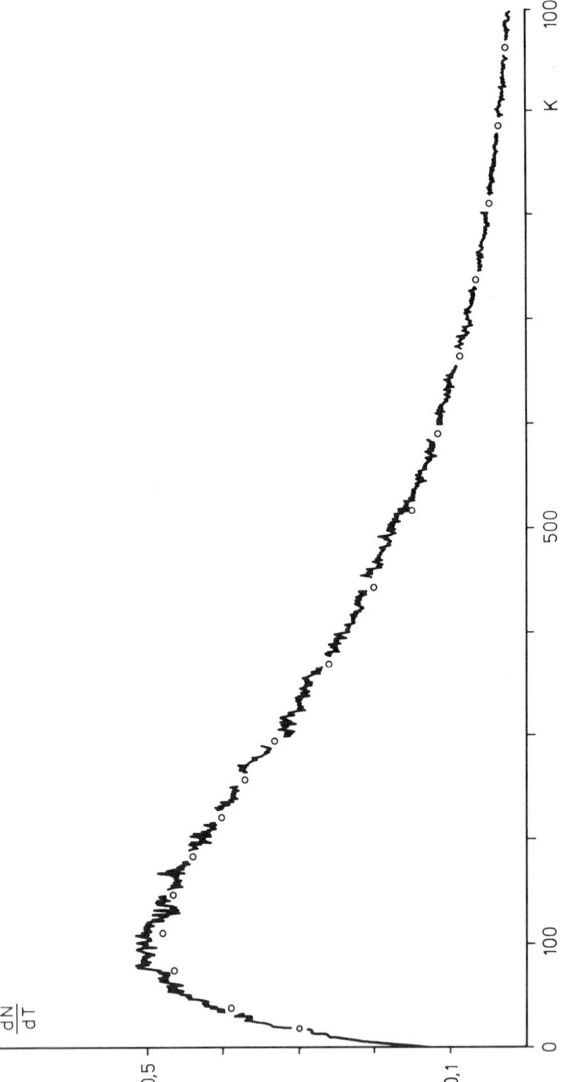

Figure 13. Comparison of the normalized rotational kinetic energy distribution of the water molecules with a Maxwellian one (dots)

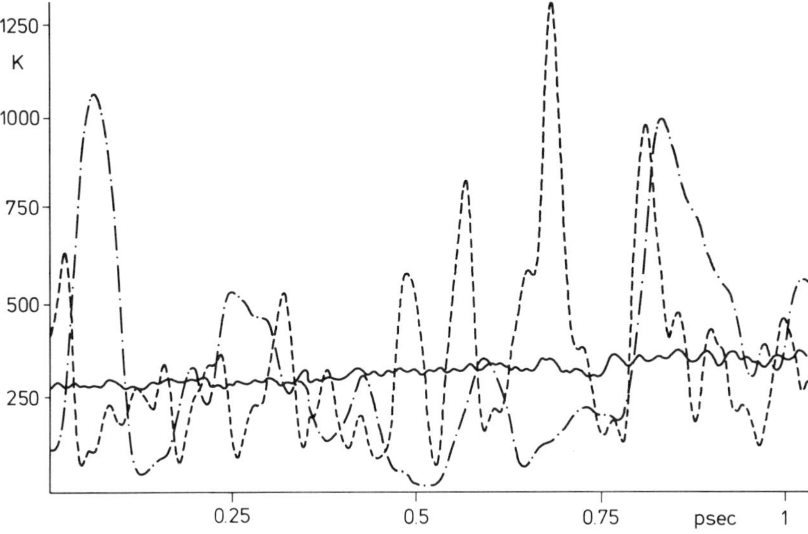

Figure 14. Kinetic energy of the total system (——), the cations (– – –), and anions (·–·–) in Kelvin as a function of simulation time

depends both on the accuracy of the programmed force routines and, as F(t) is calculated from predicted values, on the accuracy of the predicted com coordinates and Euler angles.

Simulations of aqueous solutions are characterized by the fact that the changes in the coordinates in one time step depend strongly on the type of motion. The choise of Δt for such a simulation is essentially determined by the very fast changes in the Euler angle coordinates, and Δt is thus in principle too small for the integration of the translational motion. A Δt of $1.09 \cdot 10^{-16}$ sec leads to a contribution from the corrector of $|\Delta r| < 10^{-9}$A for the com coordinates and of $|\Delta \alpha| < 10^{-6}$ for the quaternion parameters. Though, in our simulations, the accuracy of predicted values is sufficient to calculate forces, $\Delta F(t)$ may not be much larger than their inaccuracy F^{err}, leading to the above described phenomenon. This fact requires an improvement of the force routine, for example through the removal of the cut-off inaccuracies. An improved calculation of forces would allow a further step in saving computer time. Provided that a suitable extrapolation for the forces is found, the slowly varying com coordinates could be processed with a much larger step.

The evaluation of the Euler angles α, β, γ from the components of the angular velocity in a body-fixed system of coordinates leads to further difficulties. The expressions for $\dot{\alpha}$, $\dot{\beta}$ and $\dot{\gamma}$

$$
\begin{aligned}
\omega_x &= -\dot{\alpha} \sin\beta \cos\gamma + \dot{\beta} \sin\gamma , \\
\omega_y &= \dot{\alpha} \sin\beta \sin\gamma + \dot{\beta} \cos\gamma , \\
\omega_z &= \dot{\alpha} \cos\beta + \dot{\gamma} ,
\end{aligned}
\tag{33}
$$

(for the definition of the Euler angles compare (19)) become singular if $\sin\beta = 0$. Two solutions of this problem are possible. In the first case, the Euler angles are redefined, when $\sin\beta$ approaches 0 (12). This method is used for the simulation of the alcali halide solutions.

The second method is to use the so called quaternion parameters (qp) (18). These are defined in terms of the Euler angles as

$$
\begin{aligned}
\chi &= \cos\beta/2 \cos(\alpha+\gamma)/2 \\
\eta &= -\sin\beta/2 \sin(\alpha-\gamma)/2 \\
\xi &= -\sin\beta/2 \cos(\alpha-\gamma)/2 \\
\zeta &= \cos\beta/2 \sin(\alpha+\gamma)/2
\end{aligned}
\tag{34}
$$

with the auxiliary condition:

$$\chi^2 + \eta^2 + \xi^2 + \zeta^2 = 1$$

and lead to singularity-free equations instead of eq. (33). This type of parametrisation is connected with the spin-1/2 representation of the rotation group $D(\alpha, \beta, \gamma)$, as can be seen from the expressions containing half Euler angles. One easily verifies

$$D(\alpha,\beta,\gamma) = \begin{pmatrix} \cos\beta/2 \exp(i(\alpha+\gamma)/2) & \sin\beta/2 \exp(-i(\alpha-\gamma)/2) \\ -\sin\beta/2 \exp(i(\alpha-\gamma)/2) & \cos\beta/2 \exp(-i(\alpha+\gamma)/2) \end{pmatrix}$$

$$= \begin{pmatrix} \chi + i\zeta & -\xi + i\eta \\ \xi + i\eta & \chi - i\zeta \end{pmatrix} \quad (36)$$

The equations to be used instead of (33) are

$$\begin{pmatrix} \omega_x \\ \omega_y \\ \omega_z \\ 0 \end{pmatrix} = 2 \begin{pmatrix} -\zeta & \chi & \xi & -\eta \\ -\chi & -\zeta & \eta & \xi \\ \eta & -\xi & \chi & -\zeta \\ \xi & \eta & \zeta & \chi \end{pmatrix} \cdot \begin{pmatrix} \dot{\xi} \\ \dot{\eta} \\ \dot{\zeta} \\ \dot{\chi} \end{pmatrix}$$

The stability of the energy was not essentially improved after building in the quaternionic algebra, yet experience has shown that increasing the time step Δt by a factor of 2 leads to an increased upward trend in energy if the Euler angle substitution procedure is used, while the energy remains the same for the simulation using the singularity-free procedure. The latter check was preformed for an ammonium cloride solution (200 H_2O 8 NH_4^+ and 8Cl^-). In this case, the NH_4^+ molecule has a rigid tetrahedral four-point-charge distribution and also takes part in the rotational motion.

Conclusions

Some problems of numerical stabilities in molecular dynamic calculations of liquids with ionic interactions have been discussed. Mainly a sufficient high accuracy seems to be necessary for the intergration and force routine in order to get the kinetic properties of the system from long simulation runs.

In the case of aqueous solutions the main source of inaccuracy at this stage is the cut-off radius for ion-water interactions. It could be solved by the introduction of a method similar to the Ewald summation. An improvement in the intergration of the equations of motion seems to be possible by the choice of different time steps for translational and rotational motion. The pair potentials employed seem to describe aqueous solutions appropriately. It might be necessary at a later stage to introduce changes here too. Then the problem of the boundary conditions mentioned below must also be investigated. A different approach to the simulation of aquous solutions might be based on the central force model employed by Stillinger and Rahman (21) for pure water.

In simulations of ionic melts periodic boundary conditions appear which lead for the Coulomb interactions to the use of the Ewald method. It seems necessary to investigate the influence of the periodic boundary conditions and the Ewald method on the transport properties by increasing the number of particles per cube. This will also have the effect of better statistics for quantities like the pressure where the increased number of particles is more important than very long runs in time.

Acknowledgement

Financial support by Deutsche Forschungsgemeinschaft is gratefully acknowledged.

Literature Cited

1. Woodcock, L.V., Chem. Phys. Letters (1971), 10, 257.
2. Sangster, M.J.L. and Dixon, M., Advances in Physics (1976), 25, 247.
3. Heinzinger, K. and Vogel, P.C., Z. Naturforsch. (1976), 31a, 463.
4. Verlet, L., Phys. Rev. (1967), 159, 98.
5. Tosi, M.P. and Fumi, F.G., J. Phys. Chem. Solids (1964) 25, 31.
6. Ewald, P.P., Ann. Physik (1921), 64, 253.
7. Schäfer, L. and Klemm, A., Z. Naturforsch. (1976), 31a, 1068.
8. Schäfer, L., Z. Naturforsch., to be publisched.
9. Stillinger, F.H. and Rahman, A., J. Chem. Phys. (1974), 60, 1545.
10. Gear, C.W., ANL Report No. ANL-7126 (1966).
11. Nordsieck, A., Math. of Computation (1962), 16, 77-80, 22.
12. Rahman, A. and Stillinger, F.H., J. Chem. Phys. (1971), 55, 3336.
13. Hogervørst, W., Physica (1971), 51, 59, 77.
14. Pālinkás, G., Riede, W.O. and Heinzinger, K., Z. Naturforsch. (1977), 329, 1137.
15. Kong, C.L., J. Chem. Phys. (1973), 59, 2464.
16. Bopp, P., Heinzinger, K. and Jancsó, G., Z. Naturforsch. (1977), 32a, 620.
17. Heinzinger, K., Z. Naturforsch. (1976), 31a, 1073.
18. Evans, D.J., Mol. Phys. (1977), 34, 317.
19. Goldstein, H., Classical Mechanics, Addison-Wesley, Cambridge, Mass. (1953).
20. Friedman, H.L., Mol. Phys. (1975), 29, 1533.
21. Stillinger, F.H. and Rahman, A., J. Chem. Phys. (1978), 68, 666.
22. Vogel, P.C. and Heinzinger, K., Z. Naturforsch. (1975), 30a, 789.

RECEIVED August 15, 1978.

Monte Carlo Simulation of Water

C. S. PANGALI, M. RAO, and B. J. BERNE
Columbia University, New York, NY 10027

The usual Metropolis Monte Carlo(1,2) procedure when applied to the study of water at low temperatures is found to lead to bottlenecks in configuration space. In the Monte Carlo (M.C.) method different configurations of the system are sampled according to the Boltzmann distribution:

$$e^{-\beta v(\underset{\sim}{R}_1,\ldots \underset{\sim}{R}_N)} \quad d\underset{\sim}{R}_1,\ldots d\underset{\sim}{R}_N \qquad (1)$$

where $\underset{\sim}{R}_i$ is a six component vector in the case of water. Metropolis et al. devised a scheme according to which a new configuration $\underset{\sim}{R}' = (\underset{\sim}{R}_1',\ldots \underset{\sim}{R}_N')$ is generated from an old configuration $\underset{\sim}{R} = (\underset{\sim}{R}_1,\ldots \underset{\sim}{R}_N)$ by sampling a transition probability $T(\underset{\sim}{R}'|\underset{\sim}{R})$, and $\underset{\sim}{R}'$ is accepted with probability

$$P = \min[1, \frac{T(\underset{\sim}{R}|\underset{\sim}{R}')}{T(\underset{\sim}{R}'|\underset{\sim}{R})} \quad \frac{e^{-\beta v(R')}}{e^{-\beta v(R)}}] \qquad (2)$$

and rejected with probability $q = 1 - P$. Normally the configuration is changed by selecting a particle j at random, or cyclically, and displacing it from $\underset{\sim}{R}_j$ to $\underset{\sim}{R}_j'$ where $\delta\underset{\sim}{R}_j = \underset{\sim}{R}_j' - \underset{\sim}{R}_j$ is sampled uniformly between $-\frac{\Delta R_0}{2}$ and $+\frac{\Delta \underset{\sim}{R}_0}{2}$. The transition probability for an atomic fluid is therefore given by

$$T(\underset{\sim}{R}_j'|\underset{\sim}{R}_j) = \begin{cases} (\Delta x_0 \quad \Delta y_0 \quad \Delta z_0)^{-1} & \delta R_j \in D \\ 0 & \delta R_j \notin D \end{cases} \qquad (3)$$

where D defines a domain about each point $\underset{\sim}{R}_i$ in which the moves are to be sampled. Such a method when suitably modified for polyatomic fluids does not converge rapidly, particularly at low temperatures. In fact the M.C. method gives higher potential energies and lower values for the specific heat than the

0-8412-0463-2/78/47-086-029$05.00/0

molecular dynamics method for the same state(3). If we take $T(R_j'|R_j) \propto e^{-\beta v(R')}$, it is seen from Eq.(2) that p = 1. Since v(R') is not known, we expand it around R, and so

$$T(R_j'|R_j) = \begin{cases} C \; e^{-\beta \nabla_{R_j} v(R) \cdot \delta R_j} & \delta R_j \in D \\ 0 & \text{otherwise} \end{cases} \qquad (4)$$

where C is a normalization constant and ∇_{R_j} is the gradient operator with respect to positions and angles. It is easy to see that this procedure satisfies the condition of detail balance:

$$e^{-\beta v(R)} \; T(R'|R) \; P(R'|R) = e^{-\beta v(R')} \; T(R|R') \; P(R|R') \quad (5)$$

We adopted the Barker-Watts procedure(4) for performing the usual Metropolis M.C. walk. The change in orientation is accomplished by selecting the x, y or z axis at random and performing a rotation through an angle $\delta\theta$ about the chosen axis. The angle $\delta\theta$ is uniformly distributed on $(-\frac{\Delta\theta_0}{2}, \frac{\Delta\theta_0}{2})$. Our new force-bias method samples translational displacements δx_j, δy_j and δz_j according to the components of the force F_j acting on the particle j, while $\delta\theta_j$ will be sampled subject to the component of the torque along the axis selected for the rotation.

Results and Discussion

Starting from a lattice configuration with 216 H_2O molecules interacting with an ST-2 potential(5) at T = 283^{0}K, the standard Metropolis method generated a walk in which the potential energy per particle <V> fluctuated around $-130.5\varepsilon \pm 0.2\varepsilon$ ($\varepsilon = 5.31 \times 10^{-15}$ erg). Altogether nearly 500,000 moves were attempted with this method. Starting with one of the "equilibrated" states from this method, our force bias scheme generated a quite different walk. The results are summarized in Table I. Not only does the force bias method give a state with a lower mean potential energy, but the state is reproducible. Starting with a configuration generated by M.D. at 283^{0}K with V = -135.5ε, the standard scheme generates a walk in which <V> = $-135.0\varepsilon \pm 0.2\varepsilon$ whereas the force bias method yields <V> = $-134.0\varepsilon \pm 0.4\varepsilon$.

At T = 391^{0}K the two M.C. methods and the molecular dynamics method all give the same equilibrium state with statistically indistinguishable potential energies.

In the molecular dynamics method the pair force is set equal to zero when the pair distance exceeds a certain distance r_0. Thus the trajectories correspond to a shifted potential

$\phi(r) - \phi(r_0)$ where $\phi(r_0)$ depends on the orientation. The M.C. method, on the other hand deals with a truncated but unshifted potential. Thus the disagreement noted in Table I between M.D. and force bias M.C. may be ascribed to this difference between the two methods. To check out this assertion we studied a droplet of water containing 27 particles at T = 283°K, interacting with infinite range ST-2 potential. M.D. and the force bias M.C. give identical results.

In conclusion we note that Monte Carlo studies on highly structured fluids can be very misleading. At low temperatures this procedure leads to configurational bottlenecks. Although the force bias method is preferable to the standard Metropolis scheme, it is important that any sampling method be thoroughly tested to ensure that a unique equilibrium state is reached.

Table I

T = 283°K

	$\langle V \rangle^a$	Acceptance Ratio	$C_V^{\ b}$
Metropolis	-130.5±0.2	0.43	~19.0
Force bias	-133.5±0.4	0.65	~23.0
Molecular Dynamics	-137.4	----	23.8

a) In units of ε, $\varepsilon = 5.31 \times 10^{-15}$ erg.

b) In units of cal mol^{-1} K^{-1}.

Literature Cited

1. Metropolis, N., Metropolis, A. W., Rosenbluth, M. N., Teller, A. H., and Teller, E. J. Chem. Phys. (1953) 21, 1087.

2. Valleau, J. P. and Whittington, S. G., A Guide to Monte Carlo for Statistical Mechanics: 1. Highways, in:"Statistical Mechanics, Part A: Equilibrium Techniques"(Berne, B. J. ed.), Plenum Press, New York(1977).

3. Ladd, A. J. C., Mol. Phys. (1977) 33, 1039.

4. Watts, R. O., Mol. Phys. (1974) 28, 1069.

5. Stillinger, F. H., and Rahman, A., J. Chem. Phys. (1974) 60, 1545.

RECEIVED August 15, 1978.

3

Determination of the Mean Force of Two Noble Gas Atoms Dissolved in Water

C. S. PANGALI, M. RAO, and B. J. BERNE

Department of Chemistry, Columbia University, New York, NY 10027

To better understand the hydrophobic interaction between two non-polar molecules dissolved in water we have calculated the average force between two xenon atoms (Lennard-Jones Spheres) dissolved in ST2 water as a function of the separation r between the two spheres. Each xenon atom interacts with every water molecule through a Lennard-Jones (12-6) potential with parameters (σ = 3.43Å, ε = 170.1°K). The water molecules interact with each other through the familiar ST2 potential (1). Molecular dynamics simulations were performed for the two Xe atoms fixed at a separation r for several values of r ranging from 5.35Å to 9.18Å. In each of the five simulations there were 214 water molecules at a density of 1 g cm^{-3}. The temperature fluctuated around 298°K. The time step for the numerical integration was $\Delta t = 4.2 \times 10^{-16}$ sec for equilibration and half that for the final run. The initial phase point for a separation of 9.18Å was generated from equilibrated pure water by choosing two water molecules already at the desired separation, fixing their positions, turning off their coulomb interactions and adjusting their potential parameters to the Xe-H_2O parameters. Further equilibration is carried out for 2,000 Δt. This was followed for 3,000 Δt during which the average force on each Xe atom was computed. After this run at a pair separation of 9.18Å, the pair distance was reduced and the system was equilibrated. In this way, runs were produced for the other pair distances studied.

Two Xe atoms labeled a and b, dissolved in water will be distributed according to the pair correlation function, $g_{ab}(r)$,

$$g_{ab}(r) = \exp - [w_{ab}(r)]$$

where $w_{ab}(r)$ is the potential of mean force. The average force along the line joining the Xe atoms is given by

$$- \nabla_{\underset{\sim}{r}} w_{ab}(r) = \langle \underset{\sim}{F}_a \rangle - \langle \underset{\sim}{F}_b \rangle$$

0-8412-0463-2/78/47-086-032$05.00/0

where $<\underset{\sim}{F}_a>$ and $<\underset{\sim}{F}_b>$ are the average forces on the two Xe atoms determined in the molecular dynamics experiment. This can be expressed as,

$$-\frac{\partial w_{ab}(\underset{\sim}{r})}{\partial r} = F_{LJ}(\underset{\sim}{r}) + <F^{solv}(r)>$$

where $F_{LJ}(r)$ is the contribution due to the direct Lennard-Jones interaction and $<F^{solv}(r)>$ is the force between the atoms induced by the solvent molecules.

In table 1, we report how the solvent induced force $<F^{solv}>$ varies with r. The errors are estimated from the magnitude of the force components perpendicular to the line joining a and b. Strictly speaking, these should be zero. Clearly much longer runs are necessary to improve the accuracy of the work reported here. We found it difficult to determine $w_{ab}(r)$ from $<F(r)>$ because at the largest separation studied, the force is quite substantial, making the task of integrating $<F(r)>$ impossible.

The interesting feature of our results is the oscillation in $<F(r)>$; this indicates that when the two Xe atoms are a distance greater than 5.5Å apart, a water molecule is likely to take a position between them. This is consistent with the two Xe atoms residing in two cages with a monolayer of water between them. After completion of this work we learned that A. Geiger, et al (4) studied the hydration of two Lennard-Jones solute particles. Their results show a qualitative similarity to ours, but they do not give a quantitative result for the potential of mean force. Our work supports the theory of Pratt and Chandler (2) on the hydrophobic effect. Previous Monte Carlo work (3) along similar lines displayed no oscillation in $<F(r)>$, although in that study the authors explored values of r smaller than 7.0Å.

Table I

r(Å)	$\frac{F^{LJ}\sigma}{2\varepsilon_w}$	$\frac{<F^{solv}>\sigma_w}{2\varepsilon_w}$	$\frac{<F(r)>\sigma_w}{2\varepsilon_w}$
5.35	-7.55	19.89 ± 5.2	12.34 ± 5.2
6.07	-4.30	-20,20 ± 4.9	-24.50 ± 4.9
7.46	-1.19	3.23 ± 6.2	2.05 ± 6.2
8.30	-0.58	35.88 ± 10.2	36.46 ± 10.2
9.18	-0.29	33.74 ± 5.8	34.03 ± 5.8

σ_w and ε_w refer to the parameters for the ST2 potential.

Literature Cited

1. F. H. Stillinger, and A. Rahman, J. Chem. Phys. _60_, 1545 (1974)

2. L. Pratt and D. Chandler, J. Chem. Phys. _67_, 3683 (1977)

3. V.G. Dashevsky and G.N. Sarkisov, Mol. Phys. _27_, 1271 (1974)

4. A. Geiger, A. Rahman, and F.H. Stillinger, (Preprint)

RECEIVED August 15, 1978.

Applying the Polarization Model to the Hydrated Lithium Cation

CARL W. DAVID

Department of Chemistry, University of Connecticut, Storrs, CT 06268

Electrostatic models have a long history in chemistry. The Polarization Model has a genealogy which extends back to the work of Born and Heisenberg (1) and Rittner (2) in which simple electrostatics was applied to ionic compounds. Modern examples of such models include the recent work of Etters, Danilowicz, and Dugan (3), and, in Inorganic Chemistry, the work of Guido and Gigli (4). Since Heaviside's theorem precludes purely electrostatic stability of ionic compounds, a repulsive potential of some sort must be included to stabilize an isolated ionic molecule in the gas phase. The freedom to introduce such an *ad hoc* repulsive potential suggested to various people that even wider lattitude might be had in order to extend electrostatic models to include covalently bonded compounds. Our concern here is with water and its interactions, and we specialize to that substance now.

Early models of water, attempting to account for polymeric properties of the species using quasi-electrostatic schemes, placed imaginary negative charges in the vicinity of the oxygen atom. The latest such venture is, already, almost a classic. The Ben-Naim Stillinger(BNS) potential (5) was used in discussing liquid water up to and including direct numerical integration of Newton's equations of motion for water molecules (6). The potential was an outstanding success.

Several other potentials for the water-water interaction have recently appeared, as has a revision of the BNS potential known as ST2 (7, 8, 9, 10). In addition, several model potentials have appeared which incorporate polarization effects (11, 12). Each of these potentials seems quite capable of being

0-8412-0463-2/78/47-086-035$06.75/0

used to compute properties of macroscopic samples of water.

These electrostatic potentials for water all suffer a fatal defect, as they apply to frozen water- not ice- but water molecules with fixed (i.e., frozen) geometry. In an attempt to enlarge the class of water potentials, Lemberg and Stillinger (13) introduced a central force model which allowed "unfreezing" the water molecule. Here ad hoc atom pair interactions are introduced which, in toto, bind the water molecule together at its equilibrium geometry while simultaneously allowing water molecules to interact properly among themselves. This model was tested in a recent molecular dynamics simulation (14) quite successfully.

The central force model does not suffer from the frozen geometry defect of previous electrostatic models. The molecule vibrates properly, and even dissociates. However, it is internally unaffected by external electrostatic fields, and therefore does not fully mimic real water.

An electrostatic model of the water monomer, which included polarizability in some reasonable manner might allow for complete simulation of liquid water and of ionic solutions. The static dielectric properties of this ubiquitous solvent would, in a good model, be automatically correct, both on the microscopic and on the macroscopic level. The thermodynamic properties predicted by such a model would, except for minor calibration errors, be close to quantitative. Having no serious doubts about mimicking water at liquid densities, one can proceed to using more sophisticated models of the water molecule. The raison d'etre need no longer be justification of the methodology, and one can settle down to predictions and correlations confident that the errors in the predictions are inaccuracies in the model.

The Polarization Model

The Polarization Model regards a water molecule as a cluster of 2 protons(which repel each other) and an oxide ion(which is polarizable). The binding of the oxide ion to the protons is effected through an empirical potential function. The dipole moment of the molecule is adjusted through two special electrostatic modifier functions, $K(r)$ and $L(r)$.

The protons generate an electric field at the oxide anion which, in real electrostatics, would have the form

$$\tilde{E} = e \cdot \tilde{r} / r^3 \qquad (1$$

where e is the charge on the proton. Since the electrons of real water prevent the proton from actually setting up such a field, and since the protons would set up no field if r=0, it is reasonable to invent a shielding function which will account for these two facts, i.e., modify electrostatics so that it is applicable in this artificial model. The function chosen, K(r) has limits such that, at large r, K(r) goes to zero, thereby returning the electrostatics which is expected. However, at small r, the function K(r) acts to cut off the effect by lowering the electrostatic field so that at r=0, K(0) is chosen to obliterate the field completely. The formal definition of the function is:

$$\tilde{G} = e\, \tilde{r} / r^3 \; (1 - K(r)) \qquad (2$$

Given the effective electric field ($\tilde{G}$) at the oxide ion due to the protons, one has, if the oxide ion is polarizable, the induction of a dipole moment in the anion. The magnitude of this induced moment is

$$\tilde{\mu} = \alpha\, e\, \tilde{r} / r^3 \; (1 - K(r)) \qquad (3$$

Associated with this induced dipole moment on the oxide ion(which opposes the nuclear dipole moment) is a polarization energy

$$U_{pol} = (1/2) \; \tilde{\mu} \cdot \tilde{E} \qquad (4$$

Here, we choose to again modify normal electrostatics. Instead of using the true electric field in Equation 4, and instead of using the modified electric field of Equation 2, we invent yet another empirical function, L(r) which modifies the electric field in the energy term to account again for the intervening electrons and for the symmetry of the problem at r=0:

$$E' = e\, \tilde{r} / r^3 \quad (1 - L(r)) \qquad (5$$

so that

$$U_{pol} = (e/2) \quad \tilde{\mu} \cdot \tilde{r} / r^3 \quad (1 - L(r)) \qquad (6$$

The empirical functions K(r) and L(r) are set by forcing various observed properties of water to agree with the model. The functions are displayed in Figure 1 in the form of 1-K(r) and 1-L(r) Their calibration is carefully discussed in reference (15).

The total energy of a water molecule is given by:

$$E = e^2/r_{HH} + V(r_{O-H_1}) + V(r_{O-H_2}) + U_{pol} \qquad (7$$

where V is a potential energy function which looks, at large r, to be Coulombic, and which mimics the effect of the chemical bond in holding the protons to the oxide anion. The curvature and depth of this potential have been chosen to give good agreement with experimental facts about water. This potential function is also displayed in Figure 1 .

Equation 7 represents the internal energy of a water molecule. In an assembly of water molecules, one requires only an Oxygen-Oxygen interaction function (protons interact among themselves via pure Coulombic forces in this model) and one may proceed to apply the model to any system consisting of protons and oxide ions alone. This function is also discussed in reference (15).

The model, parameterized as required, successfully accounts for the "trans-linear" structure of the gas-phase water dimer, which is held together by an elementary hydrogen bond. It successfully predicts the order of stability of the gas-phase trimers of water(i.e., agrees with the existing quantum mechanical ordering), and appears quite capable of serving as a viable potential energy function for use in Monte Carlo and molecular dynamic simulations.

The details of the application of this model to various ions of the form:

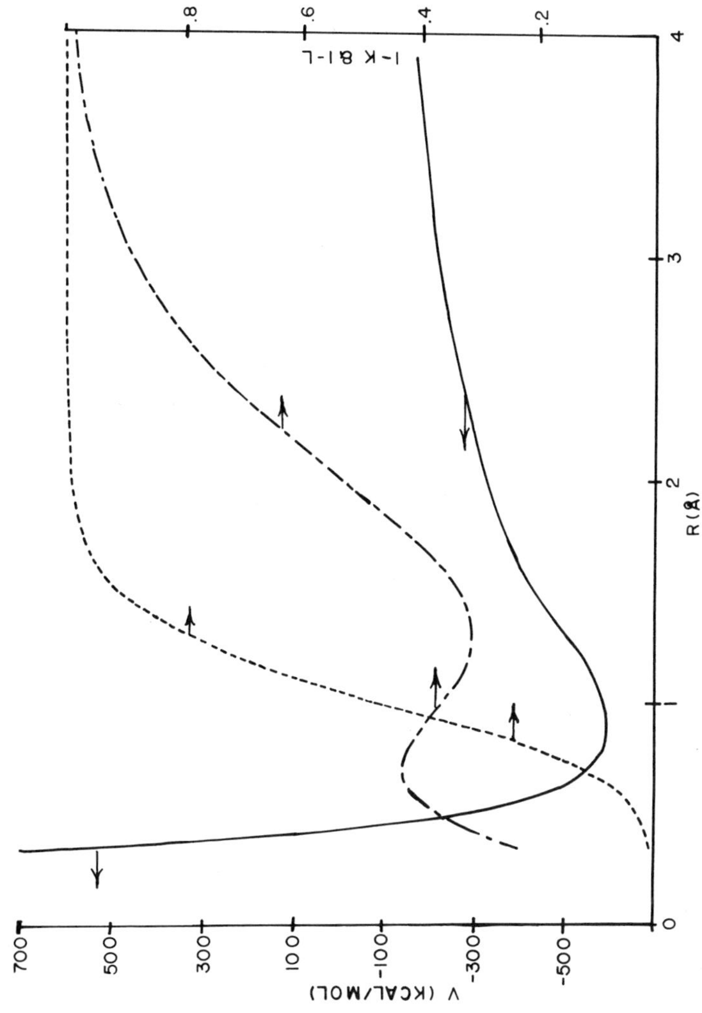

Figure 1. The calibrated functions required for the Polarization Model

$$H_mO_n^{+/-}$$

has also been given in reference (15).

We expect an almost one-to-one correspondence between the hydrated proton and the hydrated Lithium cation.

Lithium Cation-Water Potential Calibration

Since the exact potential energy of a Lithium cation in the field of a water molecule is unknown, it is convenient to assume the approximate validity of the Kistenmacher, Popkie, and Clementi(KPC) "simple" potential energy function (16). This function was invented to fit the theoretically obtained energy hyper-surface computed for "frozen" (fixed nuclear geometry)water molecules in the field of a Lithium cation. Although it is true that a "better" fitting function to the Hartree-Fock energy surface is known and reported, the simple form, which shows a minimum energy at about -35 Kcal/mol at a Lithium-Oxygen distance of 1.9 A, is adequate for our purposes.

The ground rules for parameterizing the Polarization Model for the Lithium cation (and any other species) is subject to no firm theoretical formulation. A protocol suggested by Stillinger (17) assigns K(r) and L(r) functions to species on the basis of how they appear in the formal equations of the Polarization Model. The question is moot in the cases under investigation here, as the Lithium-Oxygen and Lithium-Hydrogen distances will all be so large under normal circumstances that we can safely declare both K and L to be zero. This is equivalent to allowing their electrostatic interactions to be "classical". Pragmatically, this means employing K and L as is from the zero-order Polarization Model, as these former K and L functions are all zero at these large distances.

The value of α for the Lithium cation, required in the Model is chosen (18) to have the value

$$.03\ A^3$$

The other interactions which must be considered are the Lithium-proton interaction, and the Lithium-Oxygen interaction. The former is purely repulsive, and there is little sense in being more repulsive than the default Coulomb repulsion. We therefore set out the ground rule that we will not tinker with repulsive potentials. This leaves the Lithium-Oxygen interaction as the only interaction term with which we can adjust the interaction potential energy between the cation and water so that the total potential energy mimics (even if weakly) the theoretically derived potential functions.

Although there is no intrinsic merit to using the smallest number of parameters in fitting the potential energy, in this case, we have chosen to fit the KPC potential to the Polarization Model by employing one exponential term in the Lithium-Oxygen interaction, which also, of course, includes the Coulomb attraction of the two oppositely charged ions. This fitting function:

$$15820\ e^{-3.58\ r} \qquad (\text{kcal/mol})$$

results in a compromise fit to the KPC potential in the configuration in which the two protons are as far as possible from the Lithium ion, and the Lithium Oxygen axis is a symmetry axis.

In Figure 2 the computed potential energy function for the KPC "simple" model and the Polarization Model are displayed for the symmetric planar configuration shown. The fit is adequate. Figure 3 shows the KPC potential versus the Polarization Model potential in the reversed planar configuration indicated. The compromise nature of the "fit" is apparent.

Much to our suprise, relaxation of the structure from the planar KPC configuration results in a distinct increase in the bond strength of the complex, and, more importantly, results in a distinct tetrahedral configuration.

The absolute minimum energy of the Lithium water cation complex is -1072.28 kcal/mol, being 39.96 kcal/mol more stable than isolated water. This minimum is also to be compared to the minimum energy shown in Figure 2 which was -34.39 kcal/mol. ΔH experimentally for the process

$$Li^+ + H_2O \rightarrow Li(H_2O)^+$$

is -34.0 kcal/mol (19). Relaxation of the water has

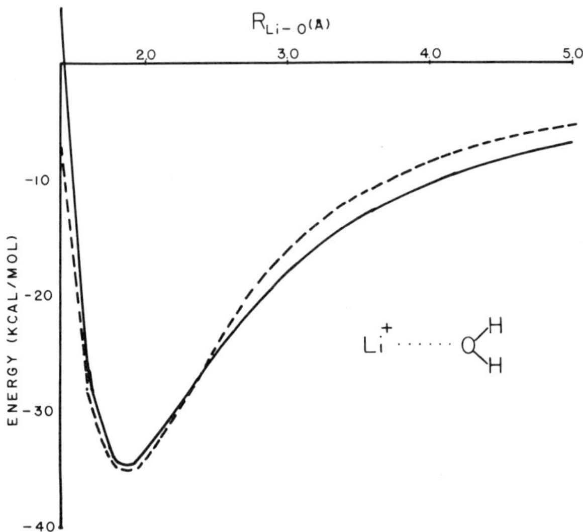

Figure 2. The Kistenmacher Popkie Clementi potential function for the lithium cation vs. water and the total Polarization Model potential for the same geometric configuration, as functions of the lithium–oxygen distance. (– – –) Polarization Model, a = .03 Å³. (——) K.P.C. potential.

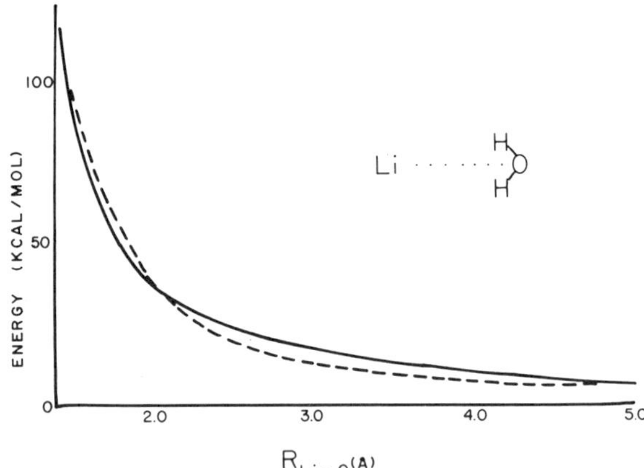

Figure 3. The same potential functions as in Figure 2 but with the configuration reversed so that the protons of the water molecule are pointing at the lithium cation. (– – –) Polarization Model, a = .03 Å³. (——) K.P.C. potential.

resulted in structure stabilization of about 5 kcal/mol. Whether or not this stabilization is real or not is open to question. We have chosen to allow this extra 5 kcal/mol stabilization to remain in the parameterization, hoping that a better quantum-chemical computation based on these suggestions will validate our assumption. (One should note that of all the ions measured by Kebarle and Dzidic, this mono-hydrate is the only one whose ΔH value was obtained by extrapolation. Therefore, more than with any of the other ions, there is a non-trivial uncertainty about the reported ΔH value of −34.4 kcal/mol.) Both O-H bond lengths in water have expanded in the complex, growing from 0.9584 A to 1.005 A. Further, the H-O-H angle has increased to 106.05 degrees. The Li-O distance has shrunk from the "fit" value of 1.9 A to 1.82 A. Using the Friedman-Lewis (20) tetrahedrality measure θ_+,

$$\theta_+ = 360. - \angle \text{Li-O-H}_1 - \angle \text{Li-O-H}_2 - \angle \text{H}_1\text{-O-H}_2$$

we find for the totally relaxed complex a θ_+ value of 17.4 degrees. In this measure, $\theta_+ = 0$ corresponds to planarity, and $\theta_+ = 31.5$ degrees corresponds to a tetrahedral configuration(where the lone-pair electrons on a real oxygen atom would be pointing directly at the central cation) Friedman and Lewis quote an experimental value of 11 degrees for the θ_+ value observed in solid

$$Li_2SO_4{:}2(H_2O)$$

They further report a bond length for the Lithium-Oxygen "bond" of 1.91 A. From the point of view of the Polarization Model, there exists no such thing as lone-pair electrons. The non-planar structure nevertheless occurs unless one assumes that the Lithium proton interaction should be made more repulsive than Coulombic.

It appears as if Vaslow (21) was the first to suggest that the dipole of a water molecule might not point directly at a central cation under circumstances such as those being considered here. This point of view did not receive theoretical corroboration from Eliezer and Krindel (22) nor does it receive such from Quantum Mechanical computations which may be found in the current literature (23-31). Spears (32) also finds the symmetric planar complex to be most stable (33).

The lack of theoretical confirmation does not preclude the possibility that the non-planar structure is, in fact, the more stable one. In their classic set of computations on ions of the series

$$H_nO_m^{+/-}$$

Newton and Ehrenson (34) obtained a minimum energy configuration for the hydronium ion which was planar, contrary to expectations. A more expensive computation by Kollman and Bender (35) was required to obtain pyramidal distortion(36). The substitution of Lithium cation for proton in the hydronium ion is not an isoelectronic replacement, and the argument by analogy need not hold. The suggestion is nevertheless painfully clear that the ion should not be planar. Since high accuracy Quantum Mechanical computations are quite expensive, one can perhaps understand why "frozen" geometry water molecules have been used in these Lithium-water complex ion computations, and therefore why planar, trigonal structures have always been found to be most favored.

There exists a modicum of experimental evidence suggesting that in concentrated LiI and LiCl solutions the Li-H distances are smaller than expected for planar structures. Geiger and Hertz (37) find that in 6m LiI solutions the Li-H distance is between 2.7 A and 2.46 A, which, assuming an Li-O distance of 2.08 A results in a canted water ligand, while Narton, Vaslow and Levy (38) find a Li-D distance of 2.43 A in 7m LiCl solution. Of course, such concentrated solutions are not equivalent to our "gaseous" ions, and the question of the structure of the ions under discussion here remains open.

Complexes of the Lithium Cation with Water.

The di-hydrate of the Lithium cation is found, using the Polarization Model as an investigative tool, to have two well defined minimum energy conformations. If one imagines the starting geometries from which relaxation leads to energy minimization, there are two simple ones which come to mind. With a water on each side of the central ion, one is the totally planar complex while the second is the non-planar structure in which one of the water ligands is canted at 90 degrees relative to the other. The difference in energy which we compute for the resultant minimum energy configurations starting from each of these two initial configurations is only 0.1 kcal/mol, clearly an insignificant difference when one realizes that the total energy is a difference between large numbers. One of these conformations is shown in Figure 4 . In both cases, the water molecules distort appreciably from their planar starting configurations (with respect to the central cation). The energy of hydration for the monohydrated cation (to form the dihydrated cation) is -30.54 kcal/mol, (experimental ΔH value = -25.8 kcal/mol) averaging over the two low energy conformers. The θ_+ value obtained for the staggered(planar) conformers was 17.32(18.84) degrees for O1, and 16.73(17.80) degrees for O2. Both water molecules have H-O-H angles of about 105.5 degrees, and all O-H distances are 0.997 A. The Lithium-Oxygen distances have increased slightly from their values in the mono-hydrate, to 1.91 A.

The higher hydrates of Lithium cation are all of interest, primarily because the enthalpy of hydration for these ions has been measured. Therefore we have undertaken to find low energy conformations of these higher hydrates, although we are well aware of the fact that the absolute minimum energy configurations are not necessarily the ones we arrive at through straight-forward minimization techniques. In fact, it is clear from our experience with hydrated protons and hydrated hydroxide ions that the hyper-surfaces in question are pock-marked with hills and valleys, and that the absolute global minimum conformation may not be accessible from all starting configurations due to algorithmic constraints. With this caveat in mind, we describe the higher hydrates of the Lithium cation.

The best, i.e., the lowest energy, conformation we can find for the tri-hydrated cation is a moderately planar one which is non-symmetric. The competition inside the first coordination sphere for hydrogen-bonding is becoming apparent, and, as a result, one of the water molecules is pushed out from the central cation so that it can accept a hydrogen bond from one of the nearer ligands. To be specific, one Oxygen is located 3.78 A from the Lithium cation, while the other two are located at 1.86 A and 1.93 A. As can be seen in Figure 5, the nearest water molecule(to the central cation), is donating a proton to the most distant water molecule. The resultant H-O-H angle for this donor molecule is smaller than normal, 103.9 degrees. The acceptor water has a normal H-O-H bond angle of 104.75 degrees. The donor molecules O-H distances are 0.99 and 1.05 A, with the long O-H bond being the donating one to the acceptor ligand. The θ_+ value for the donor ligand is 24.95 degrees(the highest value yet obtained), while the acceptor ligand has a θ_+ value of only 12.79 degrees. The lonely third ligand has a θ_+ value of 16.81 degrees and is depressingly normal in all respects.

The energy of this tri-hydrated Lithium cation is found to be -3186.6 kcal/mol, resulting in a hydration energy for the di-hydrate of -18.0 kcal/mol, compared to the experimental Δ H value of -20.7 kcal/mol.

It is not at all clear what meaning to ascribe to a structure such as that just presented for the tri-hydrated Lithium cation. To my knowledge, there is no evidence, in any physical state, to indicate whether or not this structure is correct, or reasonable, or whatever.

The last hydrated complex which will be considered in this study is the quadri-hydrated Lithium cation. After extensive(and expensive), but not exhaustive, geometric searching, we find a deepest local minimum energy configuration for this complex with energy of -4223.5 kcal/mol. This corresponds to an energy of hydration for the tri-hydrated cation of -4 kcal/mol. This is an unreasonably low value relative to the experimental value of ΔH = -16.4 kcal/mol reported by Kebarle and Dzidic. The geometry of the complex is not particularly encouraging either. As can be seen in Figure 6, the lowest energy configuration is not tetrahedral, as expected or anticipated. Rather, three of the waters are trigonally disposed around the Lithium cation, with the fourth water occupying a position roughly described as being in the second coordinate shell. The bond lengths for the various Li-O are, in their order as depicted on Figure 6, 1.91, 3.91, 2.54, and 2.04 Angstrom. The water closest to the Lithium cation is the only one participating in a hydrogen bond, (as donor), and it is bonding to the only water in the second coordination sphere. This particular donor water has a very small H-O1-H angle, 99.7 degrees. All the other H-O(n)-H angles are "normal",i.e., 103.9, 104.9 and 106.1.

The θ_+values are of some interest in this case, as the O1 θ_+= 51.21 degrees. This is, of course, exceptionally high. One must however bear in mind

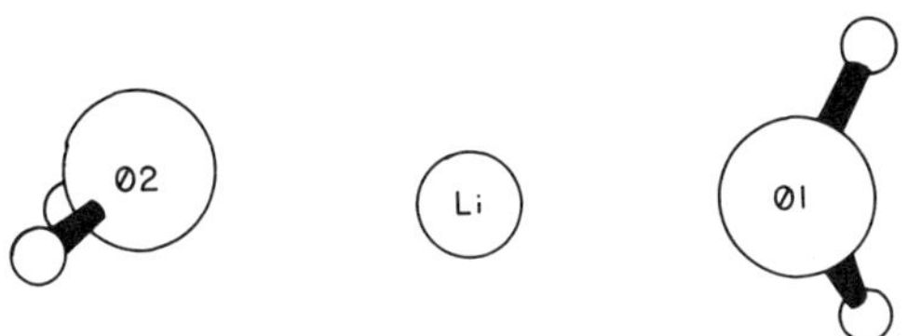

Figure 4. One of the two minimum-energy conformations found for the dihydrated lithium cation

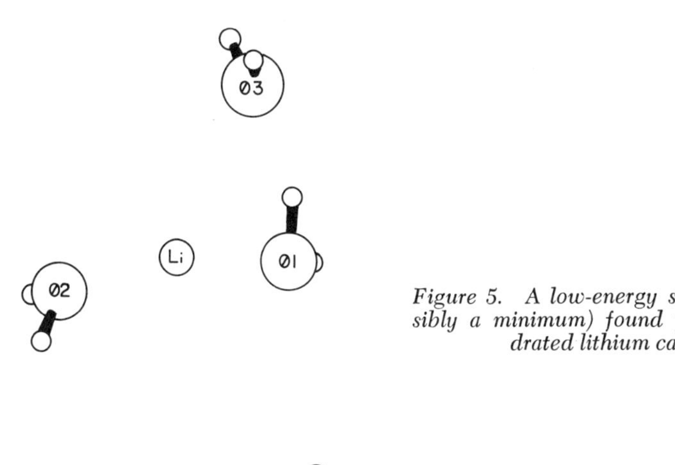

Figure 5. A low-energy structure (possibly a minimum) found for the trihydrated lithium cation

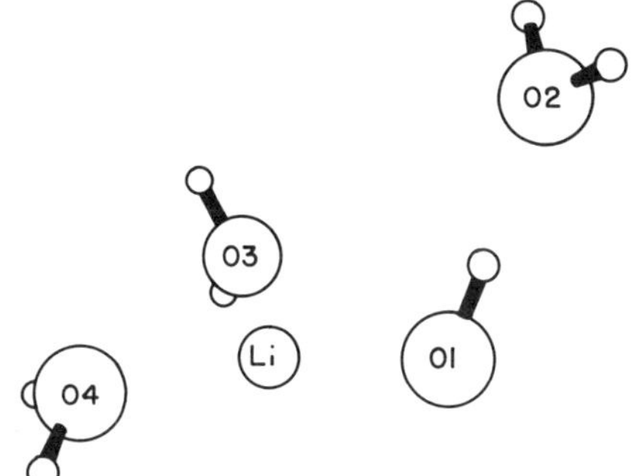

Figure 6. A low-energy structure (possibly a minimum) found for the fourhydrated lithium cation. The symmetric structure is of higher energy than this one, although there is no guarantee that the reported structure is a global minimum.

that O1 is participating in two bonds, one to the Lithium cation, and the other to another water molecule(in the second coordination sphere). The θ_+ value for O3 is closer to normal, 11.5 degrees. The θ_+ value for O4 is 20.0 degrees.

There is little doubt that this structure is not observable. We have carefully searched the nuclear configuration space in the vicinity of the symmetrically disposed tetrahedrally coordinated water complex, hoping to find a minimum energy configuration which might better correspond to the intuitively appealing pictures used by most authors in characterizing the 4-coordinated complex. This search has been fruitless, as all such symmetric configurations have corresponded to local minima which were higher in energy than the energy quoted above for the "best" structure. Nevertheless, one must regard the "true" structure of the complex as the time average of structures in which the distinction between the inner coordination shell and the outer one is blurred. Narten et alii (39) carefully refer to their symmetric structure as being a time average one.

The 4-hydrated ion is, in a sense, the solution structure, and therefore is of paramount importance. The behavior of the energy hyper-surface, described above, is not only odd in and of itself, but is at substantial variance with the equivalent behavior of the 4-hydrated proton (or the 3-hydrated hydronium ion). The Polarization Model computations on this moiety found a local minimum in the energy hyper-surface which corresponds to the roughly pyramidal structure shown in Figure 7 a and b. (We also show, for completeness, in Figure 8 the extended structure which corresponded to the absolutely lowest energy which we could find for the 4-hydrated proton. This structure corresponds closely to

$$(H_2O)_2 - H - (OH_2)_2^+$$

rather than

$$(H_2O)_3 - H_3O^+$$

Both O1 and O3 are hydrogen bond donors in this case.) The hydronium ion is based on Oxygen O3 and O3 is a "double donor" in forming hydrogen bonds. If one imagines hydrogen 6 transmuting into the Lithium cation, one sees that the ligands containing oxygens O2 and O4 are both in the second coordination sphere of the pseudo-Lithium ion. Two hydrogen bonds are being formed. Thus, the energy hyper-surface for the 4-hydrated Lithium cation is not behaving analogously to the energy hyper-surface for the 4-hydrated proton.

Although it is possible to blindly continue investigating higher hydrates, it would appear to be more profitable to turn our attention to chemical species in the Lithium, Hydrogen, and Oxygen family for which structures are actually known, even if the energies of these species are not well known.

Higher Species.

Hinchcliffe and Dobson (40 ,41) report the calculation of the structure of Li-O-Li as being "almost" planar, with a ΔH of formation of -610 kcal/mol. Using the parameterization discussed above, the minimum energy structure found had an energy of -629.3 kcal/mol, in suprisingly good agreement with the theoretical value, and a Li-O-Li angle of 153.5 degrees. The Li-O bond lengths were found to be equal, having value of 1.505 A. Unfortunately, Grow and Pitzer (42) report that Li-O-Li is linear, with a Li-O bond length of between 1.55 and 1.6 Angstrom. It is fairly clear that this linear structure is closer to the "truth" than our bent structure. We obtain a computed dipole moment for this moiety of 1.275 D, while Buchler et alii (43) report that their experimental determination places an upper bound of 0.2 D on the possible dipole moment. Of course, there is absolutely no reason that the paramaterization made on an isolated Lithium cation interacting with a water molecule's oxygen atom should be valid in the situation in which the Lithium cation is chemically bound to an oxide ion.

We have also investigated the isolated LiOH molecule, finding an optimal energy of -761.3 kcal/mol, with a Li-O-H angle of 155.5 degrees, an Li-O distance of 1.67 A and an O-H distance of .968 A. The Polarization Model computation of the energy of the hydroxide ion yields an energy for this species of -643.1 kcal/ mol, so that the binding energy of the LiOH molecule is -128.2 kcal/mol.

An occurrence of hard experimentally obtained evidence for structures in the aquated Lithium cation sequence (44) is crystallographic evidence on the structure of

$$Li(OH)(H_2O)$$

in which the Lithium ions are surrounded by two Hydroxide ions and two waters of hydration, the waters being tetrahedrally disposed. We have investigated the isolated molecule of hydrated Lithium Hydroxide, and find a minimum energy configuration with energy -1888.8 kcal/mol. Figure 9 shows the ultimate disposition of the two moieties for this isolated molecule. θ_+ for O1 is 28.1 degrees. The Li-O2-H angle is 106.5 degrees. The Li-O1 bond distance is 2.08 A, while the Li-O2 bond distance is 1.6 A. For the crystal, the Li-O(of water) distance is 1.982 A, while the Li-O (of hydroxide) distance is 1.966 A. We have also investigated the structure of this complex assuming a starting structure such as the one used in the the calculation of Urban et alii (45), where the water molecule lies between the hydroxide ion and the Lithium cation. The resultant minimum energy structure is predicted to have a higher energy, -1854.5 kcal/mol. A quantum mechanical computation using the reversed geometry would be quite interesting now.

Monte Carlo Simulations

The unsatisfactory nature of searching for minimum energy configurations for stable species which need not exist as discrete entities leads one to consider simulating larger systems. Our experience with the 4-hydrated Lithium cation demands that one investigate the behavior of this species, diluted in water, under quasi-thermodynamic conditions. This means computer simulation, either using the Monte Carlo technique or the molecular dynamics technique. Logistic considerations indicate that one should start such endeavors using Monte Carlo simulations on small systems, and we are here able to indicate a preliminary study on the 4-hydrated Lithium cation.

The Monte Carlo method is a well known scheme. As formulated by Metropolis et alii(46) a chain of configurations of the system in question is created according to an algorithm which we here describe as applied to our problem. Given a starting nuclear configuration, one generates a new possible configuration by randomly moving either a proton or an oxide ion in a random direction by a random(but

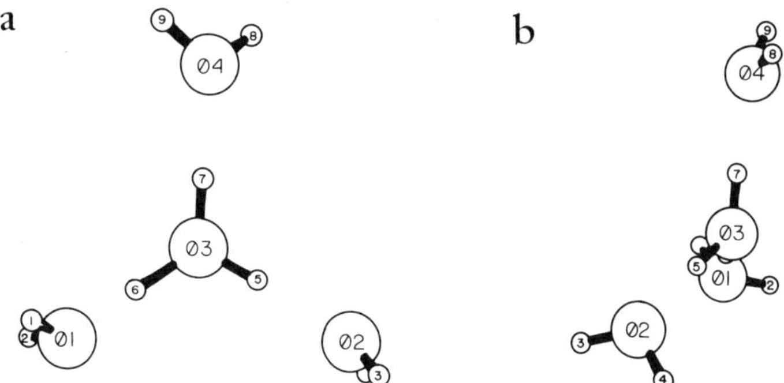

Figure 7 (a) The symmetric trihydrated hydronium ion. (b) Another view of the symmetric trihydrated hydronium ion.

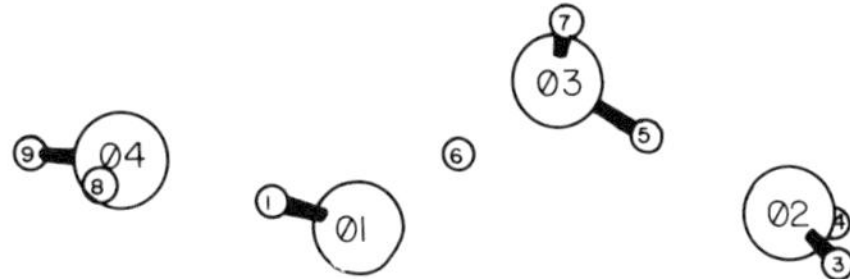

Figure 8. The lowest energy structure found for the quadrihydrated proton

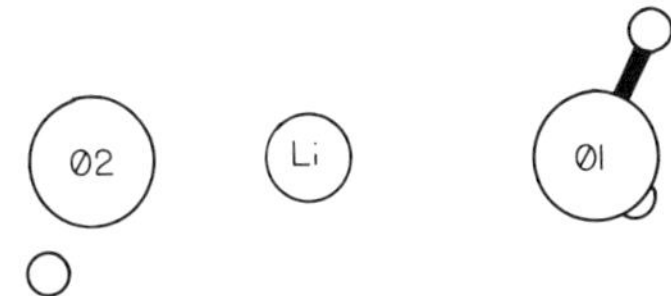

Figure 9. The hydrate of LiOH as found by the Polarization Model

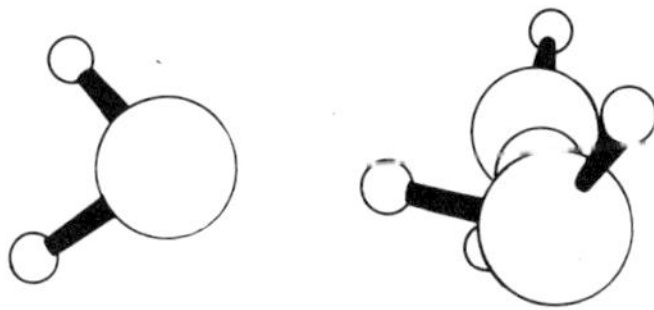

Figure 10. A snapshot of the four-hydrated lithium cation during a 300 K Monte Carlo simulation

Table 1. Pertinent Geometric Data for the Four-hydrated Proton in Two Configurations, the Symmetric One Shown in Figures 7a and 7b and the Elongated One Shown in Figure 8, and Comparable Data for the Four-hydrated Lithium Cation, Figure 6

Species	$d(O_n - H_1)$		$O_1 - O_m - O_n$	
symmetric $H_9O_4^+$	3-6	1.224	1-3-2	111.28
	1-6	1.357	1-3-4	124.40
	3-5	1.077	2-3-4	116.47
	3-7	1.076	$H_1 - O_m - H_n$	
			5-3-6	112.97
			6-3-7	123.10
			5-3-7	113.57

Species	$d(O_n - H_1)$		$O_1 - O_m - O_n$	
elongated $H_9O_4^+$	3-6	1.278	1-3-2	115.91
	1-6	1.288	1-3-4	122.71
	3-5	1.079		
	3-7	1.004	$H_1 - O_m - H_n$	
			5-3-6	114.50
			1-1-6	121.66
			5-3-7	113.57

Species	$d(O_n - Li)$		$O_1 - O_m - O_n$	
$Li^+(H_2O)_4$	3	2.039	2-1-3	90.48
	4	2.544	2-1-4	113.28
	1	1.910	1-3-4	69.02
	2	3.907	3-4-1	53.66
			4-1-3	57.32
			$Li - O_m - H_n$	
			1-1	99.75
			1-2	109.34

small) amount. The energy of this new (possible) configuration is calculated and compared to the energy of the last configuration accepted into the chain. If the new energy is lower, one chooses as the next configuration, this new one. If the new energy is higher, one forms the expression

$$e^{-(E_{new} - E_{old})/kT}$$

and compares this number to a randomly chosen number in the range 0. to 1.. (Floating point random numbers were generated using SUBROUTINE GGU3, while integer random numbers were sometimes generated from floating point random numbers by multiplication by suitable large numbers followed by truncation, and sometimes by using SUBROUTINE GGUI, both from the International Mathematical & Statistical Libraries Inc., 1977, Sixth Floor, GNB Building, 7500 Bellaire Boulevard, Houston, Texas 77036.) Should the random number be less than the expression, the new configuration is chosen to be the next one, while should the contrary occur then the old configuration is chosen again. The procedure is repeated over and over again. When stability occurs, averaging over a discrete number of configurations of the chain is carried out, with energies, heat capacities, and distribution functions being obtained (47).

It is normal to carry out Monte Carlo simulations such as that outlined above on quasi-infinite systems. This "thermodynamic" limit is carried out in two steps. First, periodic boundary conditions are invoked such that should a new configuration force a particle through one of the artificial boundary walls of the container, it disappears from the system, and a new particle appears coming through the opposite face, projecting in exactly as far as its former self projected out. We have avoided this in our simulations because the difficulty of implementation precludes rapid assessment of the technique to the problem at hand. For a central force potential model, each atom in the molecule is treated individually, and therefore, some care must be taken to make sure that periodic boundary conditions do not artificially force ionization upon the water molecule!

The second part of the passage to the thermodynamic limit involves assuming that the system being emulated is no more than a unit cell of an infinte array of replicates extending off to infinity in the three dimensional Euclidean space. It is well known that the Coulomb potential falls off too slowly to allow simple convergence of the required sum for the Madelung constant by direct numerical evaluation. An Ewald (48) sum must be carried out, and in the case of polarizable particles, a more complicated procedure must be followed (49, 50) . Since there already exists a nascent literature (51, 52) on the problem of simulation of small droplets, it seems reasonable to postpone discussion of thermodynamic systems in the context of the Polarization Model until more experience is gained with the methods of computation on the embryo droplet systems to be considered here.

As a preliminary, we have attempted to use the Monte Carlo method to ascertain whether or not the hydration number of the isolated 4-hydrated complex is 4. The Monte Carlo method is a useful adjunct to energy minimization methods for finding very low energy conformations of molecules. It is possible, by lowering the temperature, to follow the quasi-static passage of a system from a fluid state to the state of lowest energy ("solid-like").

We have run Monte Carlo simulations on the 4-hydrated Lithium cation, lowering the temperature whenever the system has appeared to come to equilibrium, until we have approached the lowest temperatures reasonable for study, ca. 50 degrees Kelvin. This is a rather unorthodox procedure in Monte Carlo work. However, it is a moderately sensitive alternative scheme to straight forward energy minimization techniques, and acts as a good cross check to guard against algorithmic constraints yielding improper minimum in the energy minimization technique. All minima, including, hopefully, the global minimum, which are connected to the starting configuration by means of only one particle moves, are theoretically accessible to the Monte Carlo minimization technique, contrary to the case of the straight forward algorthmically determined searching techniques available. Even under Monte Carlo conditions, however, no firm assurance can be made that the global minimum has indeed been discovered.

With lowering temperatures, it is required that the Monte Carlo step size be constantly readjusted to approximate a 50% acceptance ratio (when a trial configuration has an energy greater than the energy of the last configuration, a choice must be made whether or not to accept this new, higher energy, configuration. The acceptance ratio is defined as the ratio of the number of times the trial configuration is chosen compared to the total number of configurations already included in the Markov chain).

In simulating small clusters, one immediately faces a defintion problem concerning the nature of a cluster. Abraham (53) considers clusters as being bounded by a physical containing sphere (54) while Binder (55) requires instead a minimal connectedness to define a cluster (notice the rebuttal argument of Abraham (56)). We have found, as have Kaelberer and Etters (57) that at low enough temperatures, the distinction is moot, as small clusters remain intact under either definition. Therefore, we have carried out our simulations under the constraint that a Lithium cation is localized at the origin. No boundary surface is employed. In the low temperature cases, examination of the bond distribution functions shows that the clusters of hydrated Lithium cations remain intact. However, it is not at all clear that this same situation exists at higher temperatures.

In fact, the localization of the clusters provides a different problem, one discussed by Wood (58) in his famous review article. The system stays in a pocket of configuration space, without jumping to one of the N! equivalent pockets which describe interchanges of water molecules. (At the low temperatures encountered here, no ionization takes place, and therefore no interchanges of protons among waters occurs.) If we have been unfortunate enough to choose a "poor" starting configuration, one in which the ion cluster is disjoint in the sense of Binder, then, we have observed that, at high temperatures, dissociation of the cluster occurs, while at low temperatures, we observe equilibrated solid-like structure which are quite elongated. On the contrary, if we start with compact beginning clusters, then the simulation creates a chain of configurations which, in large measure, keeps the configurations compact. It is therefore difficult to know if one is in a pocket of single cluster configurations or multiple disjoint clusters. We have obtained drawings of the final clusters, and have attempted to report here only compact clusters, whose bond distribution functions indicate that they are being held together as a unit.

Results of Simulations

We have attempted to ask the question, is the

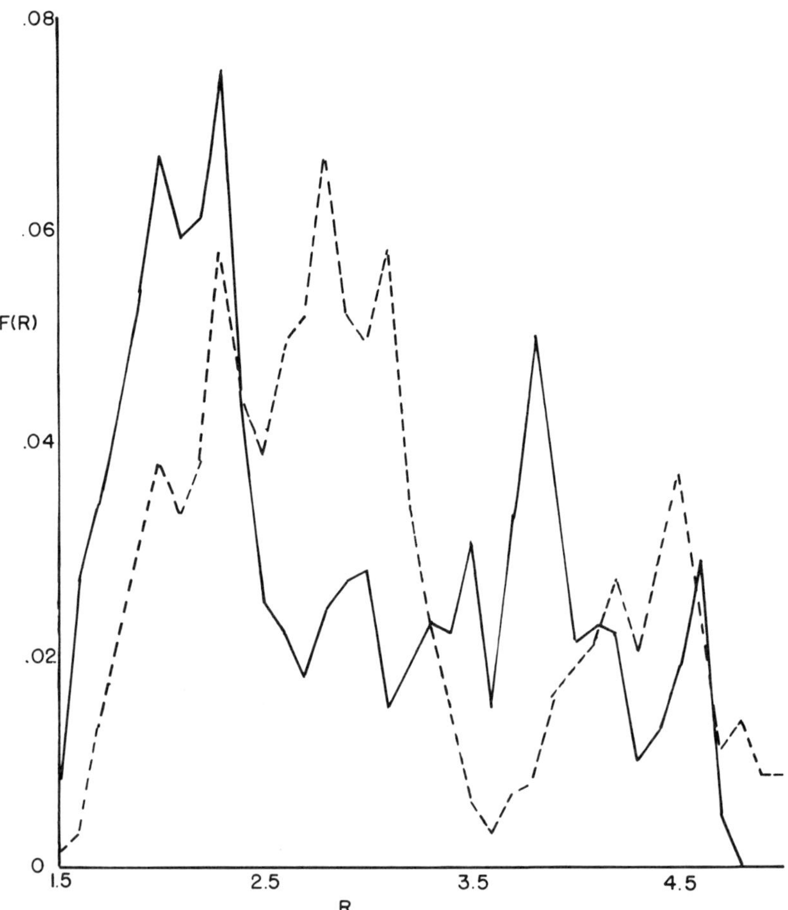

Figure 11. The bond distribution functions for the lithium–oxygen and the lithium–proton distances, taken from a 300 K Monte Carlo simulation. This function measures the fraction of named bond lengths with length between r *and* r + dr *where* dr *in our case is 0.1 Å. The Li–O function is shown in solid while the Li–H function is shown in dashed line.*

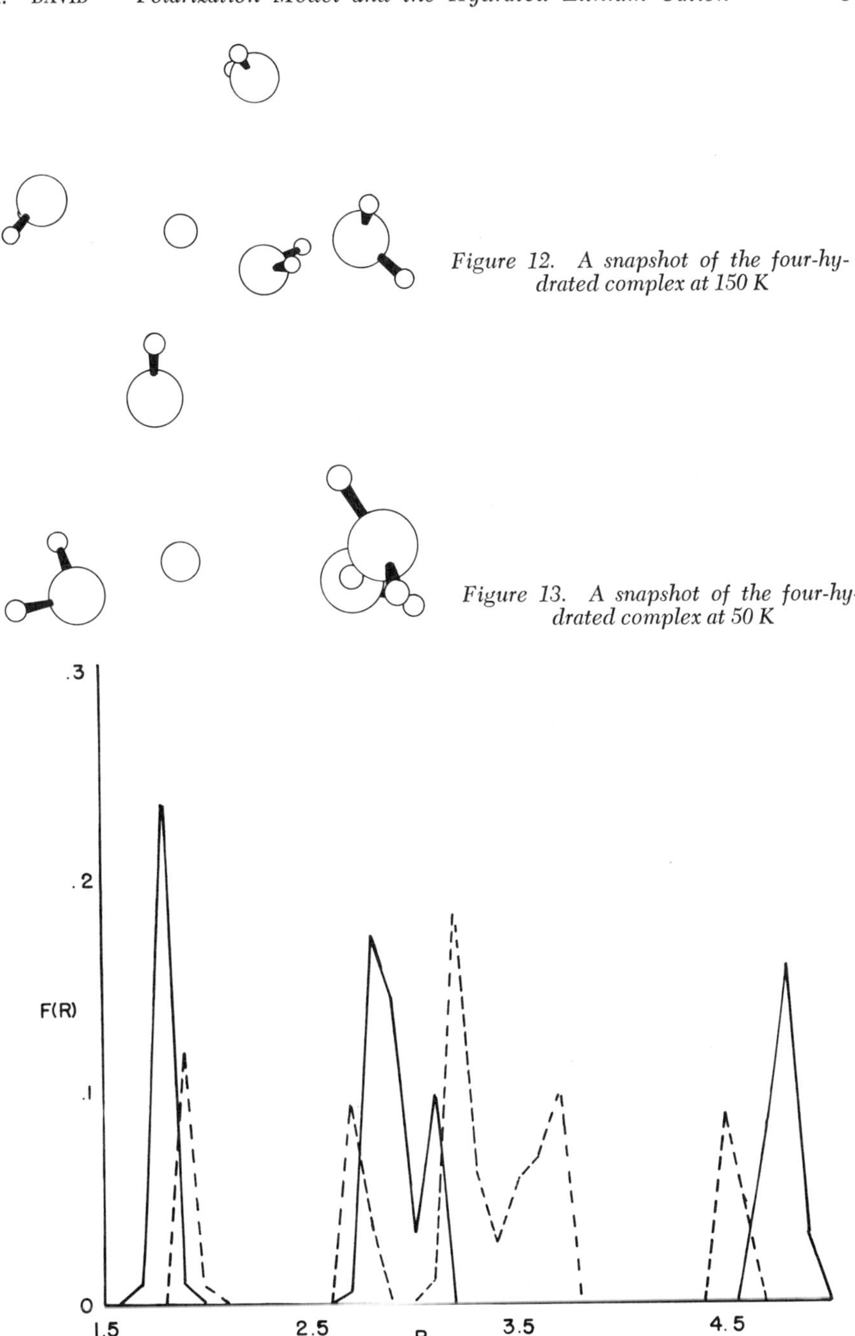

Figure 12. A snapshot of the four-hydrated complex at 150 K

Figure 13. A snapshot of the four-hydrated complex at 50 K

Figure 14. The bond distribution functions for the lithium–oxygen and the lithium–proton distances, taken from a 50 K Monte Carlo simulation

coordination number of the Lithium cation in the 4-hydrated gaseous ion 4? In order to answer this question, we have carried out a series of simulations at descending temperatures, hoping to follow the quasi-static passage to the minimum energy configuration of the complex. In Figure (10) is shown a typical configuration (in essence a snapshot) of the complex during a 300 degree Kelvin run. In Figure (11) is shown the bond distribution function for both the Li-O bond lengths and the Li-H bond lengths. The bond distribution function is defined as the fraction of all bonds of the type indicated with lengths which fall between r and r + dr. A snapshot of an intermediate simulation, Figure (12), taken at 150 degrees Kelvin shows that compactness is being achieved. In Figure (13) and (14) are shown the snapshot and bond length distribution functions for a typical 50 degree Kelvin run. It is clear that the oxygen atoms are not disposed tetrahedrally about the Lithium cation, and the coordination number of the cation is not 4. Rather, one water ligand is constantly being forced into the second coordination sphere, just as in the static energy minimizations. We are forced to conclude that no saddle point problems have occurred in the static energy minimizations which were carried out(above) on this species, and that indeed, the symmetric 4-coordination expected in this case is a phase phenomenon, resulting from packing of the hydrated ion in the dense solvent water.

A report on more traditional aspects of the Monte Carlo simulations on droplet embryos, with and without a Lithium cation present, is being prepared, and will be reported in the near future.

Acknowledgements

It is a pleasure to acknowledge the help given me by the staff of the University of Connecticut Research Computer Center, with special thanks to Peter Rutishauser and Tim Hartigan. Generous grants of computer time by the Center are gratefully acknowledged. All computations reported here were performed on the UCC IBM 360/65 & 370/155 computers.

Molecular drawings for this work were prepared using ORTEP, A Fortran Thermal-Ellipsoid Plot program for Crystal Structure Illustrations, authored by Carroll K. Johnson, Oak Ridge National Laboratory, ORNL-7 (second revision),UC.4-Chemistry,1970 ORTEP-II, March 15, 1971 version using the hidden line algorithm. Actual plotting was carried out offline using an IBM 1620 with attached plotter.

Dr. Frank Stillinger's continuing aid, encouragement, and help, is gratefully acknowledged.

Literature Cited

(1) Born M., and Heisenberg W., Z. Physik(1924)23,388.
(2) Rittner,E. S., J. Chem. Phys.(1951)19,1030.
(3) Etters R. D., Danilowicz R., and Dugan J., J. Chem. Phys.(1977)67,1570.
(4) Guido M., and Gigli G., J. Chem. Phys.(1977)66,3920.
(5) Ben-Naim A., and Stillinger F. H., "Aspects of the Statistical Mechanical Theory of Water, in Structure and Transport Processes in Water and Aqueous Solutions", Ed. R. A. Horne, Wiley-Interscience N.Y., 1972.
(6) Rahman A., and Stillinger F. H., J. Chem. Phys.(1971)55,13336.
(7) Stillinger F. H., and Rahman A., J. Chem. Phys.(1974)60,1545.
(8) Shipman L. L., and Scheraga H. A., J. Chem. Phys.(1974)78,909.
(9) Cambell E. S., and Mezei M., J. Chem. Phys.(1977)67,2338.
(10) Popkie H., Kistenmacher H., and Clementi E., J. Chem. Phys.(1973)59,1325.
(11) Campbell, E. S., and Mezei, M., J. Chem. Phys.(1977)67,2338.
(12) Berendsen, H. J. C., and van der Velde, G. A., "Molecular Dynamics and Monte Carlo Simulations on Water", Report on the C. E. C. A. M. Workshop, held in Orsay, 1972, pg. 63.
(13) Lemberg H. L., and Stillinger F. H., J. Chem. Phys.(1975)62,1677.
(14) Rahman A., Stillinger F. H., and Lemberg H., J. Chem. Phys.(1975)63,5223.
(15) Stillinger, F. H., and David C. W., J. Chem. Phys. manuscript in preparation.
(16) Kistenmacher H., Popkie H., and Clementi E., J. Chem. Phys.(1973)59,5842.
(17) Stillinger F. H., private communication.
(18) Kittel C., "Introduction to Solid State Physics", John Wiley and Sons, Inc., New York,1953,pg 165.
(19) Dzidic I., and Kebarle P., J. Phys. Chem.(1970)74,1466.
(20) Friedman H. L., and Lewis L., J. Sol'n

Chem. (1976) 5, 445.
(21) Vaslow F., J. Phys. Chem. (1963) 67, 2773.
(22) Eliezer I., and Krindel P., J. Chem. Phys. (1972) 57, 1884.
(23) Kistenmacher H., Popkie H., and E. Clementi E., loc. cit..
(24) Clementi E., and Popkie H., J. Chem. Phys. (1972) 57, 1077.
(25) Woodin R. L., Houle F. A., and W. A. Goddard III W. A., Chem. Phys. (1976) 14, 461.
(26) Diercksen G. H. F., and Kraemer W. P., Theo. Chim. Acta (Berl) (1975) 36, 249.
(27) Schuster P., Jakubetz W., and Marius W., "Topics in Current Chemistry" (1975) 60, 1.
(28) Kollman P. A. and Kuntz I. D., J. Am. Chem. Soc., (1972) 94, 9236.
(29) Kollman P. A. and Kuntz I. D., J. Am. Chem. Soc., (1972) 96, 4766.
(30) Kollman P. A., J. Am. Chem. Soc. (1977) 99, , 4875.
(31) Kollman P. A., Accounts of Chemical Research, (1977) 10, 365.
(32) Spears K. G., J. Chem. Phys. (1972) 57, 1850.
(33) Spears, K. G., and Kim, S. H. J. Phys. Chem. (1976) 80, 673.
(34) Newton M. D., and Ehrenson S., J. Am. Chem. Soc. (1971) 93, 497.
(35) Kollman P.A., and Bender C. F., Chem. Phys. Letters (1973) 21, 271.
(36) Lischka, H., and Dyczmons, V., Chem. Phys. Letters, (1973) 23, 167.
(37) Geiger A., and Hertz H. G., J. Sol'n. Chem. (1976) 5, 365
(38) Narten A. H., Vaslow F., and Levy H. A., J. Chem. Phys. (1973) 58, 5017.
(39) A. H. Narten, F. Vaslow, and H. A. Levy, loc. cit.
(40) Hinchliffe A., and Dobson J. C., Chem. Phys. (1975) 8, 166.
(41) Hinchliffe A., and Dobson J. C., Mol. Phys. (1974) 28, 543.
(42) Grow D. T., and Pitzer R. M., J. Chem. Phys. (1977) 67, 4019.
(43) Buchler A., Stauffler J., Klemperer W., and Wharton L., J. Chem. Phys. (1963) 39, 2299.
(44) Gennick I., and Harmon K. M., Inorg. Chem. (1975) 14, 2214, and references contained therein.
(45) Urban M., Pavlik S., Kello V., and Mardiakov J., Collection Czechoslov. Chem. Commun. (1975) 40, 587.
(46) N. Metropolis, A. W. Rosenbluth, M. N. Rosenbluth, A. H. Teller, and E. Teller, J. Chem.

Phys. (1953)21,1087.
(47) An excellent discussion of the Monte-Carlo method can be found in A. Ben-Naim, "Water and Aqueous Solutions", Plenum Press, N.Y. 1974, pg. 73.
(48) Ewald P. P.,Ann. Physik(1921)64,253.
(49) Kornfeld H., Z. Physik (1924)22,27.
(50) Vesely F. J.,J. Comp. Phys.(1977)24,361.
(51) Lee, J. K., Barker, J. A., and Abraham, F. F., J. Chem.Phys. (1973)58,3166.
(52) Briant, C. L., and Burton, J. J., J. Chem. Phys. (1975)63,3327.
(53) Abraham, F. A., J. Chem. Phys.(1974)61,1221.
(54) Lee, J. K., Barker, J. A., and Abraham, F. A.,J. Chem. Phys. (1973)58,3166.
(55) Binder, K., J. Chem. Phys.,(1975)63,2265.
(56) Abraham, F. A., and Barker, J. A., J. Chem. Phys.(1975)63,2266.
(57) Kaelberer, J. B., and Etters, R. D.,J. Chem. Phys.(1977) 66,3233.
(58) Wood, W. W., "Physics of Simple Fluids", Eds. H. N. V. Temperley, J. S. Rowlinson, and G. S. Rushbrooke, Wiley Interscience, John Wiley and Sons, Inc., New York, 1968.

RECEIVED August 15, 1978.

5

Molecular Dynamics Simulation of Methane Using a Singularity-Free Algorithm

S. MURAD and K. E. GUBBINS

School of Chemical Engineering, Cornell University, Ithaca, NY 14853

Rather extensive computer simulation studies have been reported for liquids of linear molecules over the past few years (1). Much less work has been reported for liquids of nonlinear molecules, water (2) being the only exception, although some simulation work for ammonia (3), benzene (4,5) and n-butane (6) has been published recently. Simulations of rigid polyatomics are complicated by the additional orientation variables, and also because the equations that must be solved for the rotational motion possess a singularity at $\theta = 0$ if the usual Euler angles are used to represent the orientations. This latter difficulty was overcome by Evans and Murad (7) by using quaternions in place of Euler angles. This results in a more efficient and accurate algorithm, and yields an order of magnitude improvement in energy conservation. A further increase in computing speed by a factor of three to eight may be obtained by using the multiple time step method developed by Streett *et al.* (8). The application of this method to rigid polyatomic molecules is described in another paper in this volume (9).

In this paper we use the quaternion method of Evans and Murad to study methane, and report preliminary results using a truncated and shifted site-site Lennard-Jones potential. The multiple time step method was not used in this work, principally because of storage limitations on the PDP 11/70 minicomputer used. The potential model studied involves five Lennard-Jones sites for each molecule, located on the C and H nuclei (C-H bond length 1.10 Å). Thus the pair potential is the sum of 25 interactions, 1 between C...C, 8 between C...H, and 16 between H...H sites. The potential is thus

$$u(\underline{r}_{12}\omega_1\omega_2) = \sum_{\alpha\beta}[u_{\alpha\beta}(r_{\alpha\beta}) - u_{\alpha\beta}(r_o)] \quad r \leq r_o \qquad (1)$$

$$= 0 \qquad r > r_o$$

where $\underline{r}_{12}$ is the vector from the center of molecule 1 to that of

0-8412-0463-2/78/47-086-062$05.00/0

molecule 2, $\omega_i \equiv \phi_i\theta_i\psi_i$ is the orientation of molecule i, $r_{\alpha\beta}$ is the distance between site α of molecule 1 and site β of molecule 2, r_o = 10.025 Å is the cutoff distance, and

$$u_{\alpha\beta}(r_{\alpha\beta}) = 4\varepsilon_{\alpha\beta}\left[\left(\frac{\sigma_{\alpha\beta}}{r_{\alpha\beta}}\right)^{12} - \left(\frac{\sigma_{\alpha\beta}}{r_{\alpha\beta}}\right)^{6}\right] \quad (2)$$

where $\varepsilon_{\alpha\beta}$ and $\sigma_{\alpha\beta}$ are the Lennard-Jones parameters (see Figure 1). Using this potential model, five sets of parameters ($\varepsilon_{\alpha\beta}$,$\sigma_{\alpha\beta}$) have been used to evaluate thermodynamic properties (configurational energy, pressure and specific heat). The best of these parameter sets is used in making a more detailed study of methane. This best set gives results superior to those of a recent study (10) utilizing the Williams (11) site-site exp-6 model, and is also superior to the results obtained by Hanley and Watts (12) using an isotropic m-6-8 potential model.

Method

The quaternion method has been described in detail in recent publications (7) so that we only give a brief summary of the method here.

The quaternion parameters χ, η, ξ, ζ can be defined in terms of Goldstein Euler angles (13) by

$$\left.\begin{aligned} \chi &= \cos(\theta/2)\cos(\psi+\phi)/2, \\ \eta &= \sin(\theta/2)\cos(\psi-\phi)/2, \\ \xi &= \sin(\theta/2)\sin(\psi-\phi)/2, \\ \zeta &= \cos(\theta/2)\sin(\psi+\phi)/2. \end{aligned}\right\} \quad (3)$$

From (3) it can be seen that

$$\chi^2 + \eta^2 + \xi^2 + \zeta^2 = 1 \quad (4)$$

The rotation matrix $\underline{A}$ (13) defined by

$$\underline{V}_{principal} = \underline{A} \cdot \underline{V}_{lab} \quad (5)$$

is given by

$$\underline{A} = \begin{pmatrix} -\xi^2+\eta^2-\zeta^2+\chi^2, & 2(\zeta\chi-\xi\eta), & 2(\eta\zeta+\xi\chi) \\ -2(\xi\eta+\zeta\chi), & \xi^2-\eta^2-\zeta^2+\chi^2, & 2(\eta\chi-\xi\zeta) \\ 2(\eta\zeta-\xi\chi), & -2(\xi\zeta+\eta\chi), & -\xi^2-\eta^2+\zeta^2+\chi^2 \end{pmatrix} \quad (6)$$

The principal angular velocity $\underline{\omega}_p$ is related to the quaternions by the matrix equation

$$\begin{pmatrix} \dot{\xi} \\ \dot{\eta} \\ \dot{\zeta} \\ \dot{\chi} \end{pmatrix} = \frac{1}{2} \begin{pmatrix} -\zeta, & -\chi, & \eta, & \xi \\ \chi, & -\zeta, & -\xi, & \eta \\ \xi, & \eta, & \chi, & \zeta \\ -\eta, & \xi, & -\zeta, & \chi \end{pmatrix} \begin{pmatrix} \omega_{p_x} \\ \omega_{p_y} \\ \omega_{p_z} \\ 0 \end{pmatrix} \qquad (7)$$

The inverse of (7) is simple to find since the 4 x 4 matrix is orthogonal. The equations of motion for these parameters are thus free of singularities.

A system of 108 model methane molecules was studied using standard periodic boundary conditions (1). The intermolecular forces and torques were calculated in the laboratory coordinate system. The torques were then converted to principal torques $\underline{T}_p$, using the rotation matrix (equation (6)). Principal torques and forces $\underline{F}$ were used to evaluate the time derivatives of the principal angular velocities and cartesian positions $\underline{r}$ of the molecules using the equations

$$\frac{d\omega_{p_\alpha}}{dt} = T_{p_\alpha}/I_{p_\alpha} \qquad (\alpha = x,y,z) \qquad (8)$$

$$\frac{d^2\underline{r}}{dt^2} = \underline{F}/m \qquad (9)$$

A third order predictor-corrector method (14) was used to integrate the center of mass motion (equation (9)), while a second order method was used for orientational equations of motion of the molecules (equations (8) and (7)).

All results were based on 2,000 time steps of 1.47 x 10^{-15} seconds, after equilibration. The energy and pressure were corrected for the long range potential contribution in the usual way (1), by putting $g_{\alpha\beta} = 1$ for $r_{\alpha\beta} > r_o$. The shift in the potential also affects the energy, but not the pressure or specific heat. The energy was corrected for this potential shift by writing $U = U_{MD} + U_{shift}$, where U_{MD} is the value corresponding to equations (1) and (2) and obtained in the simulation, and U_{shift} is the correction, given by

$$\frac{U_{shift}}{N} = 2\pi\rho \sum_{\alpha\beta} u_{\alpha\beta}(r_o) \int_o^{r_o} dr_{\alpha\beta} r_{\alpha\beta}^2 g_{\alpha\beta}(r_{\alpha\beta})$$

where $g_{\alpha\beta}(r_{\alpha\beta})$ is the site-site correlation function.

Energy conservation was within about 0.75% per 10,000 time steps. One time step took approximately 25 seconds for a 108 molecule system on the PDP 11/70 minicomputer.

Results

The potential of equations (1) and (2) involves the parameter set (ε_{CC}, σ_{CC}, $\varepsilon_{CC}/\varepsilon_{HH}$, σ_{CC}/σ_{HH}, η_{CH}, ζ_{CH}), where η_{CH} and ζ_{CH} give the departure from the usual combining rules for σ_{CH} and ε_{CH},

$$\sigma_{CH} = \tfrac{1}{2}\eta_{CH}(\sigma_{CC} + \sigma_{HH}) \tag{10}$$

$$\varepsilon_{CH} = \zeta_{CH}(\varepsilon_{CC}\,\varepsilon_{HH})^{1/2} \tag{11}$$

In this work we have studied the effect of varying ε_{CC}, σ_{CC}, $\varepsilon_{CC}/\varepsilon_{HH}$ and σ_{CC}/σ_{HH}. The values of η_{CH} and ζ_{CH} are kept fixed and are put equal to those given by Williams (11) in his parameter set VII.

Table I shows the Lennard-Jones parameter sets used in this work. In the first five sets ε_{CC} and σ_{CC} are kept fixed, and the ratios σ_{CC}/σ_{HH} and $\varepsilon_{CC}/\varepsilon_{HH}$ are varied from one set to the next. The parameters for set 1 were obtained by putting the Lennard-Jones $\sigma_{\alpha\beta}$ and $\varepsilon_{\alpha\beta}$ values equal to those used by Williams (11) in his parameter set VII for the exp-6 model. Parameter sets 2-5 show increases in ε_{HH} of either 10 or 20%, and decreases in σ_{HH} of either 0, 1, or 2%. In all five sets η_{CH} and ζ_{CH} were kept fixed at 0.972 and 1.132, respectively, as given by Williams parameter set VII.

For each of the 5 parameter sets given in Table I three state points were studied and the mean squared deviations between molecular dynamics and experimental results (15) were calculated for the configurational energy, pressure and specific heat. The three state points used were: (a) $T = 200$ K, $\rho = 10.00$ mol ℓ^{-1}, (b) $T = 225$ K, $\rho = 18.50$ mol ℓ^{-1}, (c) $T = 125$ K, $\rho = 25.27$ mol ℓ^{-1}. The first of these conditions is in the moderately dense supercritical region, the second is in the dense supercritical region, and the third is in the dense liquid region. Table II gives the root mean squared deviations found for each parameter set. Set 5 gave the best results. For this parameter set six additional state conditions were studied. The resulting molecular dynamics results are compared with experiment in Table III.

The molecular dynamics results shown in Table III may be used to rescale the values of ε_{CC} and σ_{CC}, keeping fixed $\varepsilon_{CC}/\varepsilon_{HH}$, σ_{CC}/σ_{HH}, η_{CH} and ζ_{CH}. The method of carrying out this rescaling is illustrated (for two state conditions) in Figure 2. For a single state condition (e.g. 1 in Figure 2) it is possible to find a set of (σ_{CC}, ε_{CC}) values that give exact agreement between molecular dynamics and experiment for the configurational

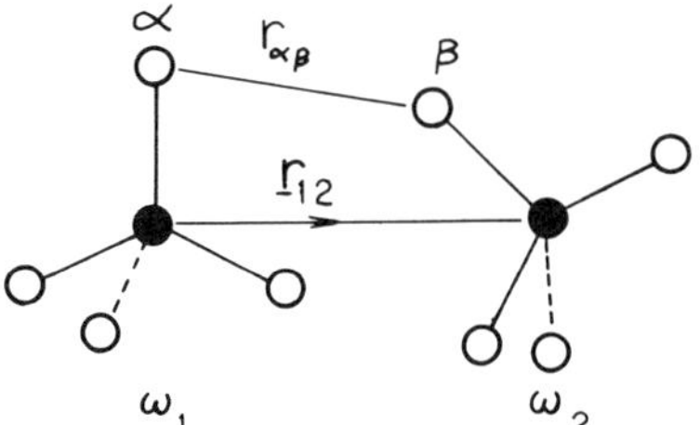

Figure 1. Site–site and center–center distances for two CH_4 molecules at orientations ω_1 and ω_2

Table I

Potential Parameter Sets Studied†

Parameter Set	σ_{CC} (Å)	σ_{CH} (Å)	σ_{HH} (Å)	ε_{CC}/k (K)	ε_{CH}/k (K)	ε_{HH}/k (K)
1	3.350	3.023	2.870	48.760	20.690	6.850
2	3.350	3.023	2.870	48.760	21.700	7.535
3	3.350	3.008	2.841	48.760	21.700	7.535
4	3.350	3.008	2.841	48.760	22.665	8.220
5	3.350	2.995	2.813	48.760	22.665	8.220
Rescaled 5	3.350	2.995	2.813	51.198	23.798	8.631

† The bond length in parameter sets 1-5 is kept fixed at 1.10 Å. The parameter set "rescaled 5" also has this same value of the bond length, since the rescaled σ's are unchanged.

Table II

Root Mean Squared Deviations For Models Studied

Parameter Set	Root Mean Squared Deviation† U^c/NkT	$P/\rho kT$	C_V^r/Nk	Total††
1	0.45	0.46	0.43	0.45
2	0.41	0.41	0.35	0.40
3	0.26	0.34	0.44	0.30
4	0.18	0.21	0.37	0.21
5	0.22	0.14	0.37	0.21
Rescaled 5	0.08	0.07	0.22	0.10

† $C_V^r = C_V - C_V^{id\ gas}$, where C_V and $C_V^{id\ gas}$ are at the same T. We expect the errors in our MD values to be no greater than 0.03 for U^c/NkT and 0.08 for $P/\rho kT$.

†† For the total deviation, energy, pressure and specific heat were given the weights 4:2:1.

Table III

Results for Parameter Set 5 Before Rescaling

Temperature (K)	Density (mol ℓ^{-1})	U^c/NkT MD	U^c/NkT Exp	$P/\rho kT$ MD	$P/\rho kT$ Exp	C_V^r/Nk MD	C_V^r/Nk Exp
198	10.00	-1.49	-1.70	0.48	0.35	0.37	0.70
228	10.00	-1.28	-1.45	0.70	0.55	0.15	0.55
250	10.00	-1.13	-1.25	0.77	0.66	0.12	0.39
223	18.50	-2.45	-2.59	0.92	0.80	0.32	0.39
248	18.50	-2.15	-2.28	1.22	1.08	0.37	0.37
276	18.50	-1.92	-1.99	1.45	1.25	0.52	0.40
123	25.27	-6.68	-6.84	-0.10	0.0	0.55	0.85
151	25.27	-5.28	-5.41	1.23	1.15	0.73	0.98
182	25.27	-4.27	-4.33	2.19	1.91	0.72	0.79

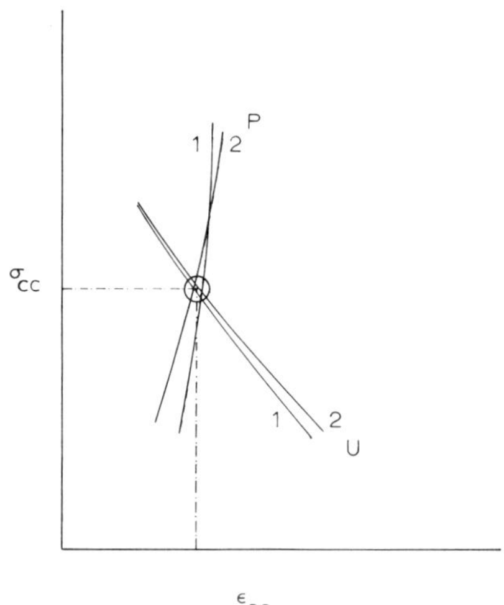

Figure 2. Rescaling the potential parameters σ_{CC} and ϵ_{CC}, keeping the ratios $\sigma_{CC}:\sigma_{CH}:\sigma_{HH}$ and $\epsilon_{CC}:\epsilon_{CH}:\epsilon_{HH}$ fixed. Curves 1 and 2 correspond to different state conditions.

Table IV

Results for Parameter Set 5 After Rescaling

Temperature (K)	Density (mol ℓ^{-1})	U^c/NkT MD	U^c/NkT Exp	$P/\rho kT$ MD	$P/\rho kT$ Exp	C_v^r/Nk MD	C_v^r/Nk Exp
208	10.00	-1.49	-1.59	0.48	0.41	0.37	0.70
239	10.00	-1.28	-1.34	0.70	0.60	0.15	0.44
263	10.00	-1.13	-1.12	0.77	0.71	0.12	0.35
234	18.50	-2.45	-2.45	0.92	0.90	0.32	0.37
260	18.50	-2.15	-2.15	1.22	1.13	0.37	0.37
290	18.50	-1.92	-1.92	1.45	1.34	0.52	0.43
129	25.27	-6.68	-6.52	-0.10	0.30	0.55	0.89
158	25.27	-5.28	-5.15	1.23	1.35	0.73	0.96
191	25.27	-4.27	-4.18	2.19	2.20	0.72	0.80

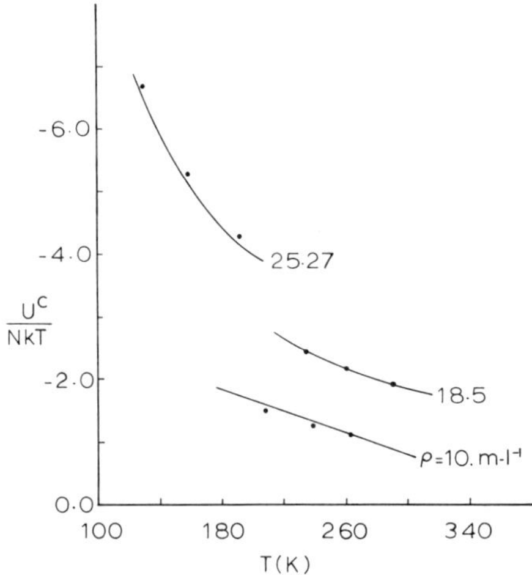

Figure 3. Configurational internal energy from molecular dynamics (points) and experiment (lines)

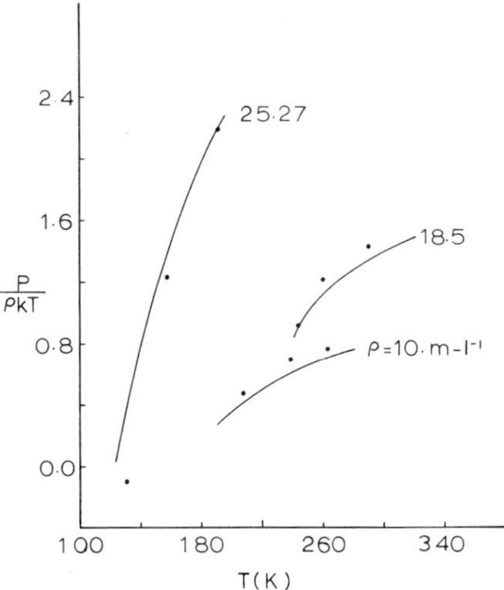

Figure 4. Pressure from molecular dynamics (points) and experiment (lines)

energy. This yields a line on a plot of σ_{CC} vs. ε_{CC}. A second, but different, set of (σ_{CC}, ε_{CC}) values is obtained by equating molecular dynamics and experimental values for the pressure at the same state condiiton. This yields a second line on the σ_{CC} vs. ε_{CC} plot. The intersection of these two lines gives a unique set of (σ_{CC}, ε_{CC}) values for the state condition studied. By treating the molecular dynamics data for the other state conditions in a similar way it is possible to estimate a best set of (σ_{CC}, ε_{CC}) values from the resulting "interval of confusion" on a σ_{CC} vs. ε_{CC} plot. This best set of rescaled parameters is given in Table I, and the resulting mean squared deviations are given in Table II. Table IV and Figures 3 and 4 compare the molecular dynamics values with experiment for this rescaled potential.

The molecular dynamics results found from parameter set 1 are almost the same as those found for the Williams potential using his set VII, and reported elsewhere (10). The results for the rescaled Lennard-Jones potential used here are substantially better, as seen from Table II. We plan to study further refinements of this potential model, including the addition of an octupole moment, and we are also in the process of investigating other properties, including the orientational correlation functions, self-diffusion coefficients, and time correlation functions.

Acknowledgments

We thank the American Gas Association, the National Science Foundation, and the Petroleum Research Fund, as administered by the American Chemical Society, for support of this work.

Literature Cited

1. Streett, W. B. and Gubbins, K. E., Ann. Rev. Phys. Chem., (1977), 28, 373.

2. Stillinger, F. H., Adv. Chem. Phys., (1975), 12, 107.

3. McDonald, I. R. and Klein, M. E., J. Chem. Phys., (1976), 64, 4790.

4. Kushick, J. and Berne, B. J., J. Chem. Phys., (1976), 64, 1362.

5. Evans, D. J. and Watts, R. O., Mol. Phys., (1976), 32, 93.

6. Ryckaert, J. P. and Bellemans, A., Chem. Phys. Lett., (1975), 30, 123.

7. Evans, D. J., Mol. Phys., (1977), 34, 317; Evans, D. J. and Murad, S., ibid., (1977), 34, 327.

8. Streett, W. B., Tildesley, D. J. and Saville, G., Mol. Phys., (1978), 35, 639.

9. Streett, W. B., Tildesley, D. J. and Saville, G., (1978), this volume.

10. Murad, S., Evans, D. J., Gubbins, K. E., Streett, W. B. and Tildesley, D. J., Mol. Phys., (1978), submitted.

11. Williams, D. E., J. Chem. Phys., (1967), 47, 4680.

12. Hanley, H. J. M. and Watts, R. O., Mol. Phys., (1975), 29, 1907; Aust. J. Phys., (1975), 28, 315.

13. Goldstein, H., "Classical Mechanics", Chapter 4, Addison-Wesley, (1971).

14. Gear, C. W., "Numerical Initial Value Problems in Ordinary Differential Equations", Prentice-Hall, Englewood Cliffs, (1971).

15. Goodwin, R. D., "The Thermodphysical Properties of Methane from 90 to 500 K at Pressures to 700 Bars", N.B.S. Technical Note No. 653 (National Bureau of Standards, Washington, DC, 1974).

RECEIVED August 15, 1978.

6

Structure of a Liquid-Vapor Interface

M. RAO and B. J. BERNE
Columbia University, New York, NY 10027

A liquid coexists with its vapor in equilibrium below the critical temperature T_c. The two well defined phases are separated by a small transition region called the interface. Recently there has been much theoretical effort devoted to an understanding of properties of the interface(1,2,3,4). It is also possible now to shed some light on the structure and dynamics of the interfaces using computer simulation(5,6,7,8). With this information one can begin to evaluate various theoretical approximations and provide a quantitative framework for the microscopic phenomenology of the liquid-vapor interface. In this paper we present some results on the liquid-vapor interface obtained from a Monte Carlo simulation.

Our model system consists of 2048 particles, interacting via a classical mechanical Lennard-Jones (6,12) potential

$$\phi(r) = V(r) - V(r_0) \qquad 0 \leq r \leq r_0$$

$$= 0 \qquad r_0 < r$$

where $V(r) = 4\varepsilon[(\frac{\sigma}{r})^{12} - (\frac{\sigma}{r})^{6}]$ and $r_0 = 2.5\sigma$. The particles are placed in a fully periodic box of size $14.7\sigma \times 14.7\sigma \times 25.1\sigma$. The details of creating an inhomogeneous system without any external field in this periodic box are presented elsewhere(5). The starting configuration thus generated is a two sided film with liquid density n_ℓ in the middle of the box and vapor density n_g on either side. Using the standard Metropolis scheme the film is equilibrated at a temperature of 110 K. A step size of 0.2σ is used for the random walk which gives an acceptance ratio of 0.5.

At equilibrium the vapor density has an average value of 0.05 and the liquid density has an average value of 0.65 (reduced units are used throughout). The symmetrized density profile is shown in figure 1 with dots. The origin is chosen to be

0-8412-0463-2/78/47-086-072$05.00/0

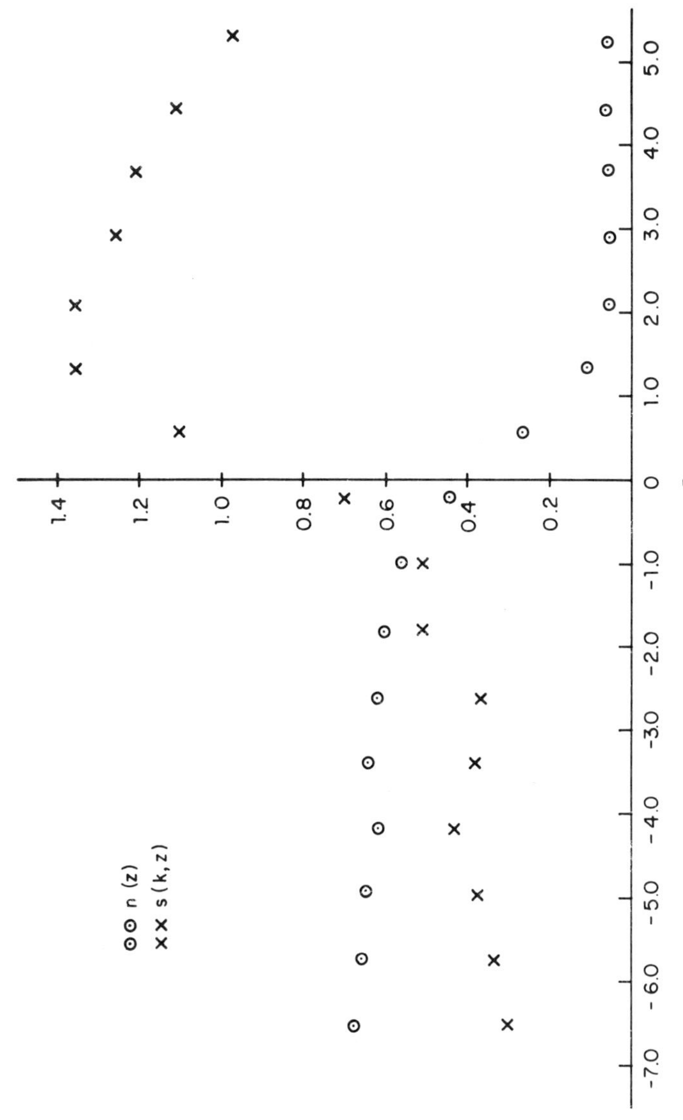

Figure 1. Density and structure factor profiles of a planar sheet of Lennard–Jonesium fluid in equilibrium with its own vapor at 110 K. The origin is chosen to be the Gibbs equimolecular dividing surface. The curves are obtained from Monte Carlo simulation using one million configurations. Circles denote the density profile and the crosses denote the structure factor, r.

the Gibbs equimolecular dividing surface given by

$$Z_g = \frac{L_Z}{2} \frac{(\bar{n} - n_g)}{(n_\ell - n_g)}$$

where L_Z is the length of the box in Z direction and $\bar{n}$ is the average density (=N/V). The two components of the pressure tensor $P_N(Z)$ (the longitudinal component) and $P_T(Z)$ (the transverse component) are determined using 1 million configurations generated during the random walk. The surface tension γ is obtained from the pressure tensor

$$\gamma = \int_{-L_{Z/2}}^{L_{Z/2}} [P_N(Z) - P_T(Z)] d Z \quad .$$

The value of γ obtained from the simulation is 0.42 reduced units. The surface tension γ_c in curved interfaces is related to the γ in a plane interface by Tolman(9)

$$\gamma_c = \gamma/(1 + 2\delta/r)$$

where r is the radius of the curved interface and δ is the curvature dependence distance. The details of estimating δ from computer simulation are presented elsewhere(10). The value obtained for δ is 1.0σ.

In figure 1 we also present the transverse structure factor $S_T(k,Z)$ previously studied(6) in a similar system at lower temperature. $S_T(k,Z)$ is defined as

$$\bar{S}_T(k,Z) = \langle \sum_{i,j=1}^{N(\Delta Z)} \exp[-ik\cdot(r_i - r_j)] \rangle / \langle N(\Delta Z) \rangle$$

where $k = (\frac{2\pi}{L}, 0)$ and $(0, \frac{2\pi}{L})$ parallel to the surface, i and j running over the $N(\Delta Z)$ atoms in a slice of volume $L \times L \times \Delta Z$ centered at Z. The enhancement of $S_T(k,Z)$ in the interface region is attributed to the capillary waves that are thermally activated. The genesis of the singular low-k transverse behavior has been discussed recently by Wertheim(2), Weeks(1) and Kalos, Percus and Rao(6). The emerging picture of the interface from these analyses is that the transition from liquid to vapor is quite abrupt in the interface -- of the order of one diameter -- but this interface fluctuates markedly in space and time. It is these fluctuations that broaden the intrinsic density profile which is quite sharp. This broadening also depends on the size of the system. Such a broadening has recently been observed in computer simulation(8).

It is now possible to simulate not only plane interfaces but also spherical droplets in equilibrium with vapor(11). The structure and dynamics of these interfaces should throw some light on the phenomenon of nucleation.

Abstract

The structure of the interface of an argon like fluid in equilibrium with its own vapor at 110 K is studied using the Monte Carlo method. The geometry of the interface is chosen to be planar and both longitudinal and transverse correlations are investigated. The longitudinal density profile shows no significant structure. From the measurement of the pressure tensor, the surface tension γ and its curvature dependence distance δ are determined. The transverse correlations exhibit very long range order in the interface suggesting a capillary wave-like behavior.

Literature Cited

1. Weeks, J. D., J. Chem. Phys. (1977) 67, 3106.

2. Wertheim, M. S., J. Chem. Phys. (1976) 65, 2337.

3. Lovett, R. A., DeHaven, P. W., Viecelli, J. J. and Buff, F. P., J. Chem. Phys. (1973) 58, 1880.

4. Singh, Y. and Abraham, F. F., J. Chem. Phys. (1977) 67, 537.

5. Rao, M. and Levesque, D., J. Chem. Phys. (1976) 65, 3233.

6. Kalos, M. H., Percus, J. K. and Rao, M., Jour. Stat. Phys. (1977) 17, 111.

7. Miyazaki, J., Barker, J. A. and Pound, G. M., J. Chem. Phys. (1976) 64, 3364.

8. Chapela, G. A., Saville, G., Thompson, G. M. and Rowlinson, J. J., J. Chem. Society, Faraday Transactions II (1977) 73, 1133.

9. Tolman, R. C., J. Chem. Phys. (1949) 17, 333.

10. Rao, M. and Berne, B. J. (to be published).

11. Rao, M., Berne, B. J. and Kalos, M. H., J. Chem. Phys. (1978) 68, 1325.

RECEIVED August 15, 1978.

7

Computer Simulation of the Liquid-Vapor Surface of Molecular Fluids

S. M. THOMPSON and K. E. GUBBINS

School of Chemical Engineering, Cornell University, Ithaca, NY 14853

The results described here are from the initial stages of a program designed to study the structure and properties of liquid-vapor surfaces of systems composed of polyatomic species. Both computer simulation and perturbation theory methods are the tools used to obtain these results. Here we report molecular dynamics (MD) computer simulations of two homonuclear diatomic liquids, each at a temperature in the region of the triple point. The liquid-vapor surface of systems of monatomic molecules has been the subject of considerable investigation (1-7) in the past few years, but as far as we are aware, this is the first work of this type on molecular species.

This work is initially directed towards obtaining both equilibrium and non-equilibrium properties in the inhomogeneous surface zone for one component systems. Equilibrium properties include the density-orientation profiles, which provides information on preferred orientations (if any) in the surface zone, and surface tensions and energies. Non-equilibrium properties include translational and re-orientational velocity autocorrelation functions and their associated memory functions, leading to information on the diffusion of molecules both perpendicular to and parallel to the plane of the interface. Here only equilibrium properties are presented. Future extensions include the study of binary mixtures of molecules of varying complexities and the behaviour of relatively massy surfactant molecules in the surface region.

I. The Simulation

Each system consists of 216 molecules confined to square prisms of dimensions x_L, y_L ($=x_L$) and z_L($>x_L$) with the usual periodic boundary conditions in the (xy) plane of the surface. The liquid was confined to the centre of the cell with vapor phases at either end. The two surfaces are stable without the inclusion of external forces (2,3,4,6). Initially the center of mass is fixed in the center ($z_{COM} = z_L/2$) of the cell. The

0-8412-0463-2/78/47-086-076$05.00/0

translational and angular velocities of the molecules are randomly chosen subject to the constraints that the centre of mass has no resultant translational or angular velocity and that the desired kinetic energy is obtained, this being divided between translation and rotational according to the equipartition law. Random orientations were assigned. The molecular centers of mass are assigned coordinates on an f.c.c. lattice structure of the appropriate liquid density, while the 'vapor phases' are 'empty'. As the simulation proceeds, the lattice structure melts and a vapor phase develops. When the resultant density profile has become stable, this initial equilibration period (typically 10^4 integration steps) is rejected and the run proper begins. A vapor molecule which attempts to exit the cell by means of one of the end walls is returned to the cell by simply bouncing it off the wall. The low vapor density ensures that this procedure results in totally negligible distortions to the surface structure. The cells were each 19 molecular diameters (σ) in height ($z_L = 19\sigma$), and the width of the cell (x_L) was computed as twice the potential cut-off distance. This latter quantity was 2.5 molecular diameters plus the molecular elongation (see below). Thus the cell widths were 5.6584 and 6.2160 molecular diameters for N_2 and Cl_2 respectively.

Each fluid was modelled by means of a site-site or atom-atom potential (8). Each molecule is assumed to consist of two interaction centers (commonly assumed positionally coincident with the atomic nuclei). The intermolecular potential is then the sum of four shifted Lennard-Jones (12,6) interactions (see figure 1)

$$u(r_{12}\omega_1\omega_2) = \sum_{K=1}^{4} u_{LJS}(r_K) \tag{1}$$

where

$$u_{LJS}(r) = u_{LJ}(r) - u_{LJ}(r_c) + u'_{LJ}(r_c)(r_c - r) \tag{2}$$

and

$$u_{LJ}(r) = 4\varepsilon[(\sigma/r)^{12} - (\sigma/r)^6] \tag{3}$$

Where r_c is the potential cut-off distance, the prime denoting differentiation, ε is the well depth and σ the molecular collision diameter of the Lennard-Jones interaction, $\omega_i = \{\theta_i\phi_i\}$ are the polar angles specifying the orientation of molecule i, and $\phi_{12} = \phi_1 - \phi_2$. The final term in (2) ensures that the potential and its derivative are continuous at the cut-off distance, leading to better dynamics. The four atom-atom distances are given by:

$$r_1^2 = r_{12}^2 + 2\ell_1^2 + 2\ell_1 r_{12}(\cos\theta_1 - \cos\theta_2) - 2\ell_1^2 f(\omega_1\omega_2)$$
$$r_2^2 = r_{12}^2 + \ell_1^2 + \ell_2^2 + 2r_{12}(\ell_1\cos\theta_1 + \ell_2\cos\theta_2) + 2\ell_1\ell_2 f(\omega_1\omega_2)$$
$$r_3^2 = r_{12}^2 + \ell_2^2 + \ell_1^2 - 2r_{12}(\ell_2\cos\theta_1 + \ell_1\cos\theta_2) + 2\ell_1\ell_2 f(\omega_1\omega_2)$$
$$r_4^2 = r_{12}^2 + 2\ell_2^2 - 2r_{12}\ell_2(\cos\theta_1 - \cos\theta_2) - 2\ell_2^2 f(\omega_1\omega_2) \quad (4)$$

with

$$f(\omega_1\omega_2) = \cos\theta_1 \cos\theta_2 + \sin\theta_1 \sin\theta_2 \cos\phi_{12} \quad (5)$$

The lengths ℓ_1 and ℓ_2 are given in terms of the atomic masses m_1 and m_2 and the bond length or elongation L as

$$\ell_1 = (m_2/M)L$$
$$\ell_2 = (m_1/M)L \quad (6)$$

where $M = m_1 + m_2$.

The potential parameters used for N_2 are those of Cheung and Powles (9)

$$\varepsilon/K_B = 37.30 \text{ K}$$
$$\sigma = 0.3310 \text{ mm} \quad (7)$$
$$L/\sigma = 0.3292$$

where K_B is Boltzmann's constant.

The parameters for Cl_2 are from the work of Singer et al. (10)

$$\varepsilon/K_B = 173.5 \text{ K}$$
$$\sigma = 0.3353 \text{ mm} \quad (8)$$
$$L/\sigma = 0.608$$

The center of mass motion is found by solving the equations

$$\ddot{\underline{r}}_i = M^{-1} \underline{F}_i \quad (9)$$

where $\underline{r}_i$ is the center of mass vector of molecule i, $\underline{F}_i$ is the force on i due to all the other molecules and the dots indicate differentiation with respect to time. A third order predictor-corrector method was used.

The rotational motion is found by solving the equations

$$\dot{\underset{\sim}{\omega}}_{ip} = I^{-1} \underset{\sim}{T}_{ip} \quad (10)$$

where $\underset{\sim}{\omega}_{ip}$ and $\underset{\sim}{T}_{ip}$ are the principal angular velocity of and torque on molecule i respectively, and I is the moment of inertia. The orientations were expressed by means of the quaternion parameters (11), and the solution of (10) in terms of these parameters is discussed in (11). The use of quaternions to represent the orientations gives singularity-free equations of rotational motion with better resultant energy conservation.

The equations of motion were integrated with a reduced time step of 0.0015 $[t^* = t(\varepsilon M/\sigma^2)^{1/2}]$. In real units this is 4.388×10^{-15} and 3.526×10^{-15} s for N_2 and Cl_2 respectively.

Total energy was not exactly conserved, but oscillated around a fixed value with a period of about 3000 time steps and an amplitude of about 0.2%. This behaviour replaces the slow drift in energy found when the intermolecular force is discontinuous at the cut-off radius, that is, the final term in (2) is omitted.

The calculations were performed on a PDP 11/70 computer. The computer programs were overlaid and occupied a maximum of 28.5 K 16-bit words of core. A time saving scheme was implemented in which molecules were ordered according to the height coordinates of their centers of mass. This was found to increase computing speed by ∿30%. Execution speed was 12.5s per step for the N_2 simulation. Configurations and translational and angular velocities were stored on disks for further analysis.

II. Properties Calculated

The surface tension γ_{sim} was calculated during the simulation by means of the equation

$$\gamma_{sim} = (N_s A)^{-1} \sum_{s=1}^{N_s} \sum_{i<j} \frac{r_{ij}^2 - 3z_{ij}^2}{2r_{ij}^2} \sum_{K=1}^{4} \frac{1}{r_K} \frac{\partial u_{LJ}(r_K)}{\partial r_K} \, \underset{\sim}{r}_{ij} \cdot \underset{\sim}{r}_K \tag{11}$$

where i and j are molecular subscripts, A is the surface area and the $\underset{\sim}{r}_K$ are implicitly assumed to refer to the molecular pair (i,j). The sum over s is over a sequence of N_s time steps. The perturbation provided by shifting the Lennard-Jones potential is allowed for in the simulation, thus making the calculations apply to a truncated diatomic Lennard-Jones fluid. The potential shift results in a shift in the coexisting liquid density (7) which is not allowed for by this method. The long-range tail of the pair potential may be approximately taken into account by adding the correction γ_{tr} , calculated as in Appendix B of (7).

The density-orientation profile $\rho(z,\theta)$ gives the probability density for finding molecule at height z with the molecular axis at an angle θ to the z axis (14). It may be written

$$\rho(z,\theta) = \sum_{\ell=0}^{\infty} \rho_\ell(z) P_\ell(\cos\theta) \tag{12}$$

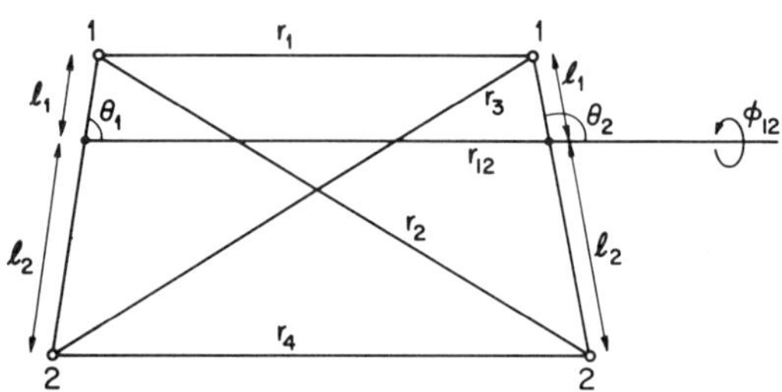

Figure 1. The atom–atom potential

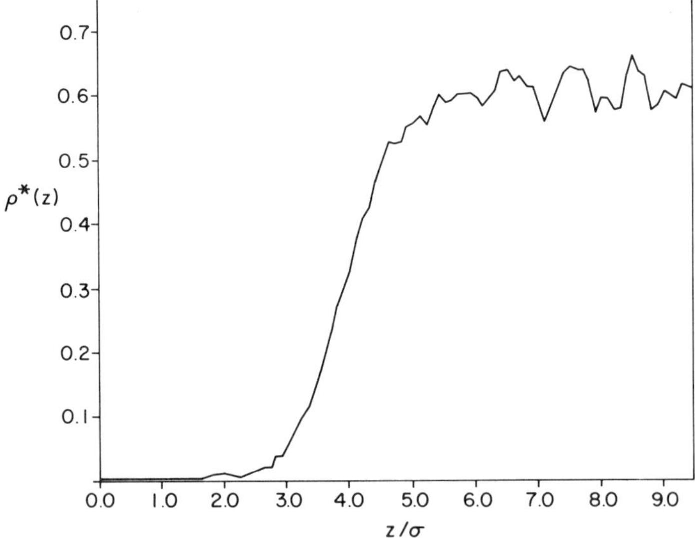

Figure 2. Total density profile for N_2 at 66.52 K. Reduced density vs. height.

where $P_\ell(x)$ is a Legendre Polynomial. The total density profile is

$$\rho_{tot}(z) = \int_0^\pi \rho(z,\theta) \sin\theta d\theta = \sum_{\ell=0}^{\infty} \rho_\ell(z) \frac{2\delta_{\ell o}}{2\ell + 1} = 2\rho_o(z) \tag{13}$$

where $\delta_{\ell o}$ is a Kronecker delta and the individual coefficients $\rho_\ell(z)$ can be calculated from the equations

$$\rho_\ell(z) = \tfrac{1}{2}(2\ell + 1) \ll P_\ell(\cos\theta_i) \gg /(0.1\ A\sigma) \tag{14}$$

where θ_i is the angle between the molecular axis of molecule i and the vertical, and the average symbolized by $\ll \gg$ in (14) is over all molecules with their centers of mass in a slice of thickness 0.1 σ (and volume 0.1 $A\sigma$) about the indicated z coordinate over the entire simulation. For homonuclear diatomic molecules, symmetry requires that all coefficients $\rho_\ell(z)$ with odd ℓ are zero. Coefficients calculated during the simulations were ℓ=0, and 4.

III. Results

Averages evaluated during the simulations are given in Table 1. The coexisting liquid densities are low as has been observed previously (7). This is due to the shift in the pair potential. The surface tensions are in good (5%) agreement with experiment, the nitrogen value being high while the chlorine value is low. The surface thicknesses are defined in terms of the gradient of the total density profile at the Gibbs dividing surface (7). The values of surface thickness shown in Table 1 differ significantly from the values for a monatomic fluid under equivalent conditions (7).

The total density profiles are plotted in Figures 2 and 3 for N_2 and Cl_2 respectively. These are the averages for the two surfaces in the box in each case; the profiles for the two surfaces were found to be identical within statistical error. The profiles for N_2 and Cl_2 are similar in their major details: the 'structure' observable in the liquid region changed in form during the simulation and is not considered significant. The real-time duration of the current simulations is substantially less than that of the simulations of monatomic species published to date because of the need to consider rotational motion, leading to higher estimated error in the liquid densities. In Figure 4 is plotted the second density harmonic coefficient $\rho_2(z)$ for Cl_2 at 172 K, and in Figure 5 the density-orientation profile at the z values indicated by * in Figure 4. The non-zero value in the bulk liquid is statistical noise: the coefficient here differed substantially in both box halves and also oscillated as the simulation progressed, while the peaks at

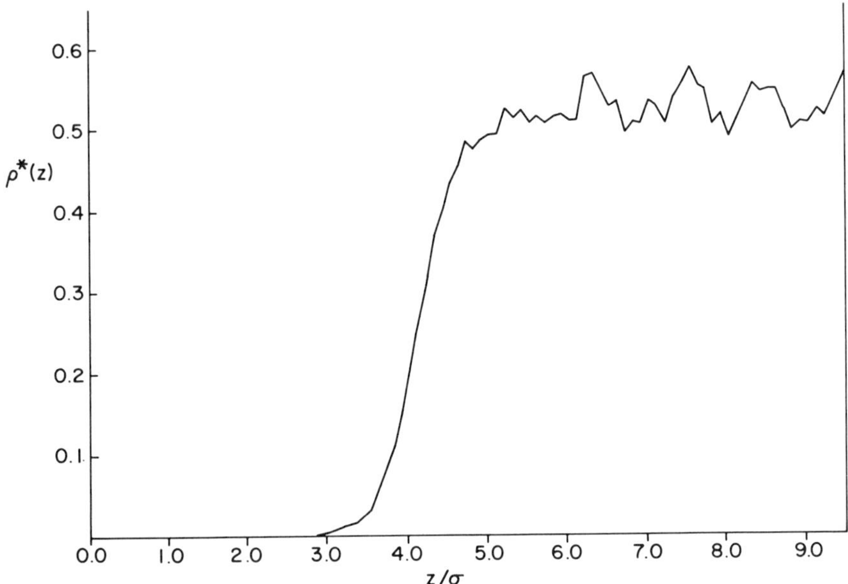

Figure 3. Total density profile for Cl_2 at 172.0 K

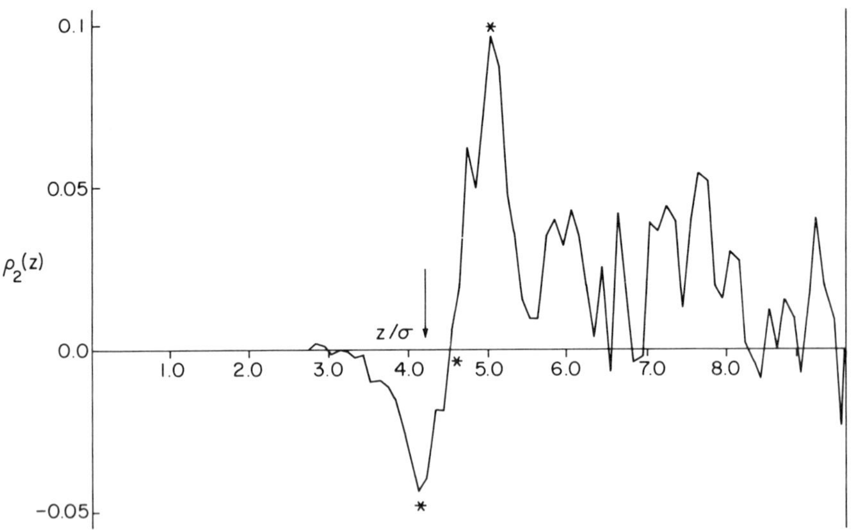

Figure 4. Second density harmonic coefficient $\rho_2(z)$ for Cl_2 at 172.0 K. The arrow locates the Gibbs dividing surface.

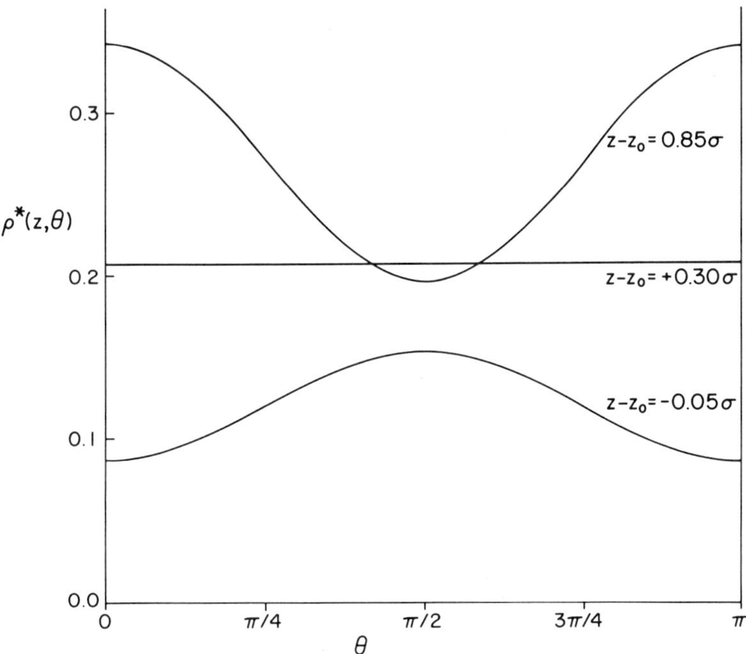

Figure 5. Density-orientation profile $\rho_0(z) + \rho_2(z)P_2(\cos\theta)$ for Cl_2 at 172.0 K

Table I Averages Obtained During Simulations

	N_2	Cl_2
Number of time steps	42.75 x 10^3	40.00 x 10^3
Temperature, K	66.52 ± 2.01	172.0 ± 5.0
Liquid density, $\rho^*_{L, sim}$	0.608 ± 0.024	0.526 ± 0.023
Surface thickness, d/σ	1.79	1.14
Location of Gibbs dividing surface, z_0/σ	4.20	4.00
Surface tension, $Jm^{-2} \times 10^{+3}$	11.98 ± 0.44	37.35 ± 0.83
Experimental coexisting liquid density, $\rho^*_{L,exp}$ (12)	0.664 at 66.5 K	0.55 at 172.0 K
Experimental surface tension, $Jm^{-2} \times 10^3$ (13)	11.41 at 66.5 K	39.06 at 172.0 K

$z \sim 4\sigma$ and $z \sim 5\sigma$ is genuine structure, being free of these features. The corresponding curve for N_2 is not plotted because of poor statistics for this near-spherical molecule. A pronounced tendency to adopt preferred orientations is indicated, this tendency being height dependent. In the liquid phase at $z - z_0 = 0.85\,\sigma$ (corresponding to the maximum in $\rho_2(z)$) the molecules have a tendency to orient with their axes vertical ($\theta = 0$), while at the Gibbs surface (close to the minimum in $\rho_2(z)$) a reversal takes place and the molecules bend to orient with their axes parallel to the interfacial plane. This result is qualitatively in agreement with the predictions of first order perturbation theory for anisotropic overlap forces (14). A perturbation treatment using the full atom-atom potential is under way.

The pair potential is currently being modified by the inclusion of a quadrupole-quadrupole potential, and the effect of surface area on surface thickness found in (7) for monatomic molecules is being investigated.

Acknowledgment

It is a pleasure to thank the National Science Foundation and the Petroleum Research Fund, administered by the American Chemical Society, for grants in support of this work, and Sohail Murad for a copy of his simulation program for hydrogen chloride.

Abstract

An application of the molecular dynamics method to simulate the liquid-vapor surface of molecular fluids is described. A predictor-corrector algorithm is used to solve the equations of translational and rotational motion, where the orientations of molecules are expressed in quaternions. The method is illustrated with simulations of 216 homonuclear (N_2 and Cl_2) diatomic molecules. Properties calculated include surface tensions and density-orientation profiles.

Literature Cited

1. Chapela, G. A., Saville, G. and Rowlinson, J. S., Faraday Disc. Chem. Soc. (1975) 59, 22.
2. Lee, J. K., Barker, J. A. and Pound, G. M., J. Chem. Phys. (1974) 60, 1976.
3. Abraham, F. F., Schreiber, D. E. and Barker, J. A., J. Chem. Phys. (1975) 62, 1958.
4. Opitz, A. C. L., Phys. Letters A (1974) 47, 439.
5. Liu, K. S., J. Chem. Phys. (1974) 60, 4226.
6. Rao, M. and Levesque, D., J. Chem. Phys. (1976) 65, 3233.
7. Chapela, G. A., Saville, G., Thompson, S. M. and Rowlinson, J. S., Trans. Far. Soc. II (1977) 73, 1133.
8. Sweet, J. R. and Steele, W. A., J. Chem. Phys. (1967) 47, 3029.
9. Cheung, P. S. Y. and Powles, J. G., Mol. Phys. (1975) 30, 921.
10. Singer, K., Taylor, A. and Singer, J. V. L., Mol. Phys. (1977) 33, 1757.
11. Evans, D. J. and Murad, S., Mol. Phys. (1977) 34, 327.
12. Vargaftik, N. B., Tables on the Thermodphysical Properties of Liquids and Gases, 2nd ed., Halsted Press Div., Wiley, New York (1975).
13. Jasper, J. J., J. Phys. Chem. Ref. Data (1972) 1, 841.
14. Haile, J. M., Gubbins, K. E. and Gray, C. G., J. Chem. Phys. (1976) 64, 1852.

RECEIVED August 15, 1978.

8

High Field Conductivity

BENSON R. SUNDHEIM

Department of Chemistry, New York University, 4 Washington Place, New York, NY 10003

In the computation of transport properties from molecular dynamics data, both direct and indirect (fluctuation-regression) means have been used. The indirect method deals with systems at equilibrium whereas the direct methods do not.

Here we study the transport properties associated with electrical fields by the examination of the steady-state properties of a simulated fused salt exposed to a uniform field. The latter is large by laboratory standards in order to produce statistically useful displacements. This means that the upper portion of the linear response is explored in a way that is not readily accomplished in the laboratory. Since large fields imply that there must be substantial heat dissipation, it is necessary to "thermostat" the system so that, as potential energy is withdrawn from the electric field, kinetic energy is withdrawn from the system. By monitoring this withdrawal, an alternative determination of the conductance can be made. Several interesting byproducts of this computer "experiment" are discussed below, including the rate at which kinetic energy in one degree of freedom is "thermalized" into others, the properties of the fluctuating dipole moment per unit volume and the relation between experiments in the laboratory and the computer results. A technical modification in the means of computations has led to the use of a relatively large (40 Å) cell so that the relatively long wave length propagating modes can be explored.

The molecular dynamics calculation of electrical conductivity in ionic fluids has been approached in several different ways. The autocorrelation function of the current may be related to the conductivity by

0-8412-0463-2/78/47-086-086$05.00/0

Kubo type relations (1). By expansion it may be separated into terms containing only the autocorrelation function of a given particle and terms contain cross-correlation functions. The latter can be identified as being the source of deviations from the Nernst-Einstein equation.(2)

A second, quite novel method, is to treat the electric field as producing a small perturbation to the particle trajectories (3). The molecular dynamics calculation is carried out twice, once without the field and once with the field applied as a step function or as an impulse function. By comparison of the two sets of calculations, it is possible to obtain an estimate of the electrical conductivity. The accuracy and, indeed, the fundamental justification for the method has not yet been well established.

Finally, there is a "brute force" method (4). Remembering that the motion produced by an external electrical field is a small perturbation on the Brownian motion, it can be seen that very high fields can be applied without sensibly altering the gross properties. In computer experiments, millions of volts per centimeter can be applied without concern for electrode processes and thermostatting can be supplied to prevent significant temperature changes. Consequently, it is possible to examine the range of applicability of Ohm's Law and the details of the transfer of energy from the field to one component of the kinetic energy and hence to all three degrees of freedom and ultimately into the thermostatic bath. The conductivity can be determined both by determining the rate at which heat is extracted from the system and by determining the particle mobility in the applied field. It is this high field method which is the first subject of this communication.

Computational Details

For convenience, the potential energy of interaction, aside from the coulombic term, was chosen to be the same for all particles, being a rough approximation to that appropriate to molten KCl. Its form was that recommended by Woodcock (5). The symmetry of the potential plus the absence of ionic polarization terms means that the results are not specific to any real salt but rather pertain only to this model. On the other hand, the main results are meant to refer to the properties associated with dense ionic fluids in general and not to unique properties associated with various anomalies in the potential energy of interaction.

The coulombic contribution was evaluated by the Ewald method (5,6). The direct force contribution was evaluated for a pair of particles by using a pre-calculated force/distance table based on the potential energy function. Pairs with distances (either direct or to the nearest image) greater than 6.6 A were omitted by use of a "link" system (7) as adapted to fused salt systems by Woodcock (8). The applied electrostatic field was simulated by adding a constant force term (of opposite sign for the two ionic species) to the direct force on each particle. This results in a flow of energy into the system. Compensatory thermostatting was applied at each time step (10^{-14} sec) by determining the ratio of the net kinetic energy to that required by the nominal temperature and dividing each velocity by the square root of this ratio. The Verlet algorithm (9) was utilized for the integration. Autocorrelation functions were computed by the Fourier transform method.(10).

The rescaling of the velocities in order to maintain the desired rms velocity in each cartesian coordinate is an approximation to thermostating.A better procedure would be to scale in such a way as to maintain a Boltzman distribution. The linear scale factor utilized here represents the leading term in a power series expansion in such a Boltzman rescaling. Noting that the average correction coefficient is less than 0.998 even for very high fields, it may be readily seen that the relative deviation between the distribution achieved and a Boltzman distribution is quite small and unimportant in these studies. The rescaling plays no role in the momentum balance so that the system may be described as a constant temperature, constant potential gradient, constant momentum, constant density ensemble. (The density is constant over a spatial "graining" of the order of the cell volume, here approximately 40x40x40 A, and the temperature is constant over the temporal graining of 10^{-14} sec.)

Experimental measurements of electrical conductivity in real systems ignore heating effects wherever possible, or remove them by extrapolation to zero applied field. In our computations a similar extrapolation to zero applied field was made.

Results

We turn now to an examination of the results of the computations which illustrate the method.

The average current flow over 500 steps in systems which had previously come to a steady state was used to calculate the electrical conductivity. The results are summarized in Fig.1. A slight apparent positive slope to the least squares line is, in fact, well within the 95% confidence level of zero slope (dashed line) and we conclude that up to 15 megavolts per cm the isothermal conductivity is constant for this model. (Further decreases in the uncertainty of the slope, obtainable by longer runs and/or more points, did not seem to be especially valuable.) The underlying reason for the constancy of the conductivity is, of course, the fact that the external force is only a tiny fraction of the rms fluctuating force experienced by a particle.

There are two other means of obtaining the electrical conductivity from the same computation. The amount of heat extracted from the system at each step was recorded and divided by the square of the current. The resulting quantity is also the conductivity. Values obtained in this way are also shown in Fig.1 (+'s) . Finally the power spectrum of the electrical current in a field free system was computed and the autocorrelation function derived therefrom used to obtain the electrical conductivity via the well-known relationship (1,2). The value so obtained is entered on Fig.1 as an open circle. All of these methods are, in principle, equivalent. The highest accuracy is attached to the direct measurement of the mean current.

The diffusion coefficient can be obtained by the limiting slope of the graph of the mean square displacement per particle versus time. Alternatively, the velocity autocorrelation function may be utilized for the calculation (1,2).The statistical reliability of the velocity autocorrelation function is quite high for this large system since it represents the mean of the autocorrelation function for 1728 particles over several sets of 500 steps.

Comments

There are two points peculiar to the computation of electrical conductivity that should be discussed here. One has to do with the mode of the electrostatic current waves and the other with dielectric shielding.

Both transverse and longitudinal current oscillations occur in the molecular dynamics computations (6). The periodic boundary conditions mean that the

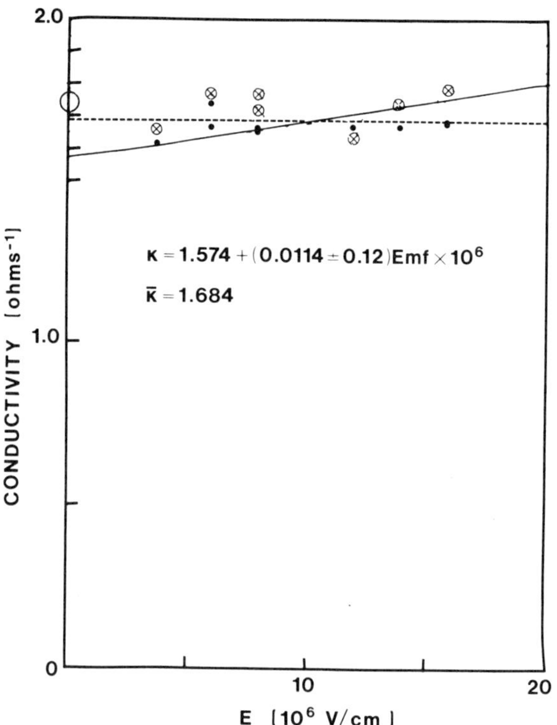

Figure 1. Computed conductivity as a function of applied field. The solid line represents the least squares fit to the high-field method points and the dashed line the mean value of these points. The open circle is the zero field point obtained by the current correlation function method. The crosses are the points obtained by the heat dissipation method.

current will be divergenceless for wavelengths equal to or greater than the dimensions of the elementary cell. When the DC current is obtained by averaging over the cell, longitudinal currents are eliminated. (The zero k vector wave does not occur since the periodic boundary conditions imply a "toroidal" space which is also divergenceless at + infinity and hence has no planes of charge accumulation.) These calculations therefore refer to a zero k vector transverse wave, which is the quantity measured on a real salt with reversible electrodes and direct current.

The dielectric properties are classically pictured in terms of "free" and "bound" charges associated with an electrical fluid. In a fused salt system, charge is carried by both types of particles and, for non-polar particles, there are no charges which can be bound. Nevertheless, it is possible to polarize the system of ions by inducing a charge separation. In the simplest terms, an external field biases the local distribution functions, giving rise to a reaction field which modifies the mean fluctuating force on each particle. We may cast this in the conventional form by the following argument.

The linear expression for the current-voltage relation is (11)

$$\vec{J} = \sigma\vec{E} + \chi(\partial\vec{E}/\partial t)$$

In an isotropic fluid σ and χ are scalars. The polarization per elementary cell of edge S is

$$\vec{P} = \sum_i z_i(\vec{r}_i - \vec{n}_i S)$$

where $\vec{n}_i$ is the (integral) number of displacements through the unit cell wall experienced by a particle which was originally in the cell centered at (0,0,0). Then

$$(\partial\vec{P}/\partial t) = \sum_i z_i\vec{v}_i - S\sum_i(\partial\vec{n}_i/\partial t)$$

Since $\sum z_i\vec{v}_i = \vec{J}$,

the net current observed by averaging over the cell and the "mobile charge current", j may be written as

$$\vec{j} = S\sum(\partial\vec{n}_i/\partial t)$$

Then $\vec{J} = \vec{j} + (\partial\vec{P}/\partial t)$
Thus, $(\partial P/\partial t)$ plays the role of the "bound charge current". Of course, in the steady state J = j. By comparison, since

$$\vec{j} = \sigma\vec{E}$$
$$(\partial P/\partial t) = \chi(\partial E/\partial t) = \partial(E\chi)/\partial t$$
$$\vec{P} = \chi\vec{E}$$

by setting

$$E = D - 4P\pi$$

$$D = \varepsilon E$$

so that

$$\varepsilon = 1 + 4\pi\chi$$

We are thus led to the idea that the current obtained by averaging the charge-weighted velocities over the unit cell and that obtained by monitoring the net flux through the surface of the cell may not be the same in dynamic situations. Each of these quantities is separately found in a molecular dynamics computation and may be used in a fluctuation dissipation calculation of the individual contributions to the conductivity and hence to the polarizability of the medium. The applied emf is, of course, equivalent to an external field in the absence of surface binding dipoles.

One last point: It is well known that the real and imaginary parts of the electrical conductivity are connected by the Kramers-Kronig dispersion relations. (11) The relation between the two is generally obtained by contour integration with due care being given to poles. The fused salt system, not being an inertialess fluid, does not have an instantaneous response to a voltage pulse and so its complex conductivity may be taken to be analytic in the upper half-plane. It is therefore possible to obtain the imaginary part of the frequency dependent conductance (at zero k vector) of this model by numerically inverting the Fourier transform of the real part and then recomputing the transform, obtaining both the real and imaginary parts. Numerical computations dealing with this technique and with the polarizability are in progress. (12)

References

1. a. Green,M.S., J. Chem. Phys., 22, 398 (1954).
   b. Kubo, R., J. Phys. Soc. Japan, 12, 570 (1957).
   c. Kubo, R., Rep. Prog. Phys., 29, 255 (1966).
   d. Zwanzig, R., Ann. Rev. Phys. Chem., 16, 67 (1967).
2. Hansen, J. P. and McDonald, E.L., J. Chem. Phys., 62, 4581 (1975).
3. Gicotti, G. and Jacucci, G., Phys. Rev. Lett., 31, 206 (1973).
4. cf. the application of this method to viscosity by
   a. Ashurst, W. T. and Wainwright, T. E., J. Chem. Phys., 53, 3813 (1970).
   b. Gosling, E. M., Singer, I. R. and Singer, K., Mol. Phys., 26, 1475 (1973).
5. e.g., Woodcock, E. L., p.8 in Advances in Molten Salt Chemistry, Vol. 3, ed. Braunstein, J. Mamontov, G. and Smith, G. P., Plenum Press, New
6. Hansen, J.P. and McDonald, E.L.,Theory of Simple Liquids, Acad. Press, New York (1976).
7. a. Schofield, P., Comput. Phys. Comm., 5, 17 1973).
   b. Quentrec, B. and Brot, C., J. Comput. Phys., 13, 430 (1973).
8. Woodcock, L.E., private communication.
9. Verlet, L., Phys. Rev., 159, 98 (1967).
10. Futrelle, R.P. and McGinty, D. J., Chem. Phys. Lett., 12, 785 (1971).
11. Landau, L.D. and Lifshitz, E.M., Vol. 5, p 391 ff., Pergamon Press, New York (1958).
12. Valuable conversations with Dr. L.E. Woodcock (Cambridge), grants of computer time from the Courant Institute, N.Y.U. under D.O.E grant E(11-1)-3077 and support from the National Science Foundation are gratefully acknowledged.

RECEIVED August 15, 1978.

9

Computer Simulation of Collective Modes in Solids

M. L. KLEIN

Chemistry Division, National Research Council of Canada, Ottawa, Canada K1A 0R6

The molecular dynamics technique is now widely used to calculate the timedependent properties of many body systems. This paper reviews the application of this computer simulation technique to the calculation of the normal modes of vibration of solids and their lifetimes. Examples will be representative of most classes of solid found in nature, namely the inert gases, ionic crystals, metals, alloys, and molecular solids.

Normal modes of vibration of solids are routinely studied experimentally using inelastic scattering of neutrons or photons (1). Also the theory of the dynamics of crystals is now well developed and even anharmonic effects are usually incorporated into theoretical models (1). This being the case one may wonder why computer simulation is at all necessary? Well, neutron scattering is rather costly so that although this technique is in principal very powerful it is limited in this sense and light scattering is mostly restricted to small wave vectors. The simulation method, on the other hand, is really a compliment to the real experiments since it allows one to compare approximate theoretical approaches to the crystal dynamics with exact results for a given intermolecular force model, subject only to the limitation that classical statistical mechanics applies. Fortunately, for most solids, this latter restriction is not a severe limitation. Moreover, in the computer simulations there is an unambiguous distinction between coherent and incoherent scattering, and between one-phonon and multiphonon effects, a situation which does not always pertain to real experiments. We have used the molecular dynamics (MD) method originally pioneered by Alder and Wainwright (2) for hard spheres and by Rahman (3) for continuous potentials. The collective modes (phonons) we wish to study are related to the spectrum of density fluctuations in the solid. Such motions are obtained by integrating the classical equations of motion (using a time step of 10^{-14} to 10^{-15} secs) for a model system composed of approximately 10^3 particles, interacting via some assumed force law, with the periodic boundary condition being used to simulate an infinite system. These boundary

0-8412-0463-2/78/47-086-094$05.00/0

conditions "quantize" the allowed wave vectors that can be studied. The details of the phase space trajectories are stored so that at a later time statistical averages can be performed. This in essence is the method. It has been widely discussed in the literature and will not be described in detail here.

The Dynamical Structure Factor $S(\underset{\sim}{Q},\omega)$

It is well known that apart from dull factors the coherent scattering of neutrons from liquids and solids is determined by the Van Hove function (dynamical structure factor)

$$S(\underset{\sim}{Q},\omega) = \int_{-\infty}^{\infty} dt e^{i\omega t} F(\underset{\sim}{Q},t)$$

where $F(\underset{\sim}{Q},t)$ is the intermediate scattering function. This in turn is related to the time correlation function of the density operator

$$\rho_{\underset{\sim}{Q}}(t) = \sum_{i=1}^{N} e^{i\underset{\sim}{Q}\cdot\underset{\sim}{r}_i(t)}$$

where N is the number of particles in the system whose position at time t are given by $\underset{\sim}{r}(t)$. The momentum and energy transferred to the system in the scattering process are $\underset{\sim}{Q}$ and ω respectively. In detail

$$F(\underset{\sim}{Q},t) = (1/N)\langle\rho_{\underset{\sim}{Q}}(t)\rho_{-\underset{\sim}{Q}}(0)\rangle ,$$

where the angular bracket denotes a statistical average which will be evaluated by the classical molecular dynamics method (MD).

The first calculations of this type were carried out for liquid rare gases near the triple point (4). Since then a variety of fluids have been studied including liquid metals (5), liquid nitrogen (6), the classical one-component plasma (7), molten salts (8) and water (9). Because liquids are isotropic $S(\underset{\sim}{Q},\omega)$ depends only on $|\underset{\sim}{Q}|$ whereas for solids this is not true. A further distinction arises because for many solids under a wide variety of state conditions the constituent particles execute small amplitude vibrations about well defined equilibrium positions (noteworthy exceptions are plastic crystals). If this situation pertains it is possible to express $\underset{\sim}{r}_\ell(t) = \underset{\sim}{R}_\ell + \underset{\sim}{u}_\ell(t)$ the instantaneous position of particle ℓ in terms of its mean position $\underset{\sim}{R}_\ell$ and its time dependent displacement $\underset{\sim}{u}_\ell(t)$. If $\underset{\sim}{u}_\ell$ is in some sense small with respect to $\underset{\sim}{R}_\ell$ one can develop $F(\underset{\sim}{Q},t)$ as a power series in $\underset{\sim}{Q}\cdot\underset{\sim}{u}$. This is the so-called phonon expansion (10)

$$F(\tilde{Q},t) = F_0 + F_1 + F_{int} + F_2 + \cdots$$

where the successive terms describe the elastic (zero-phonon) scattering one-phonon inelastic scattering, the interference between one and two phonon scattering, and the two-phonon scattering, etc. For a classical solid these individual correlation functions can be evaluated using the MD method as can the full $F(\tilde{Q},t)$.

From a technical point of view one can proceed as follows. One can evaluate $F(\tilde{Q},t)$ for a given $\tilde{Q}$ and then take the Fourier transform. Alternatively, one can proceed via the Fourier-Laplace transform of the density operator $\rho_{\tilde{Q}}(\omega)$ using the formula

$$N\,S(\tilde{Q},\omega) = \lim_{\tau\to\infty} \int_0^\tau e^{i\omega t}\rho_{\tilde{Q}}(t)dt \int_0^\tau e^{-i\omega t'}\rho_{-\tilde{Q}}(t')dt' / \tau$$

$$= \lim_{\tau\to\infty} |\rho_{\tilde{Q}}(\omega)|^2 / \tau$$

Both methods have been used in the literature (11). Other methods involving perturbations of trajectories have also been used to study collective modes in solids (12).

The one-phonon approximation to the dynamical structure factor S_1 can be written (13)

$$N\,S_1(\tilde{Q},\omega) = \lim_{\tau\to\infty} |\tilde{\rho}_{\tilde{Q}}(\omega)|^2 / \tau$$

where

$$\tilde{\rho}_{\tilde{Q}}(t) = e^{-W} \sum_{\ell=1}^{N} e^{i\tilde{Q}\cdot\tilde{R}_\ell}\, \tilde{Q}\cdot\tilde{u}_\ell(t)$$

and

$$W \cong \frac{1}{6}|\tilde{Q}|^2\langle u^2\rangle$$

is the exponent of the Debye-Waller factor.

Since this is the quantity most usually calculated by approximate theoretical models (10) it is of considerable interest to evaluate $S_1(\tilde{Q},\omega)$ and to compare this with the full $S(\tilde{Q},\omega)$ for a given model. In the following sections we consider in turn the application of the MD technique to Rare Gas Solids (13), Alkali Halides (14), Metals (15,16) and Alloys (17) and Molecular Crystals (18,19).

Rare Gas Solids

The rare gas solids usually crystalline in a fcc structure. The interatomic forces are predominantly pairwise additive but there is a considerable body of evidence to suggest that qualitative models require the presence of 3-body contributions (20). Nevertheless the Mie-Lennard-Jones (12-6) potential provides a good effective model with which to study the rare gases.

$$v(R) = 4\varepsilon[(\sigma/R)^{12}-(\sigma/R)^{6}]$$

A convenient unit of temperature is $T^* = k_B T/\varepsilon$ and density $\rho^* = \rho\sigma^3$, where for solid Ar $\varepsilon/k_B \simeq 120$ K and $\sigma \simeq 3.40$ Å. Extensive calculations of $S(\underset{\sim}{Q},\omega)$ for the (12-6) potential were carried out by Hansen and Klein (13). They carried out systematic studies of the temperature and density dependence of the response function. Typical spectra are shown in Figure 1. The unit of energy corresponds to 18 cm^{-1}, 0.54 THz, or 2.2 meV for ^{36}Ar. The upper half of the figure corresponds to approximately one half the melting point of solid Ar while the lower curve is close to the triple point. The full lines are the X-point response which includes both longitudinal and traverse phonons for the zone boundary <001> direction. The dashed dot curve shows the longitudinal response alone. In general the high frequency phonons were very broad at high temperatures and this explains why they have proved difficult to measure experimentally.

One novel feature of the simulations (13) was the presence of a central Rayleigh peak for the smallest wave vector $\underset{\sim}{Q} = (2\pi/a)(1/8,0,0)$ studied at high temperatures in the system of 2048 particles. The width of this peak is governed by the thermal diffusivity and provides a crude estimate of the thermal conductivity since the specific heat is known from other calculations.

Alkali Halides

The first MD study of alkali halides utilised the rigid ion model. This venerable old model consists of point charges plus Born-Mayer repulsions and dispersion force attractions. The limitations of this model are well known to be due principally to neglect of polarization effects. The subject of the interatomic forces has been reviewed recently by Sangster and Dixon (21). Jacucci et al. (14) studied NaCl at several temperatures using the MD method. A system of 216 ions was used and the Coulombic interactions were handled by an Ewald method (21). Results for the smallest wave vector studied at 302 K are shown in Figure 2. The insert figure shows the LO (longitudinal Optic) phonon whereas the main peak corresponds to the LA (Longitudinal Accoustic) mode. Figure 3 shows the same phonon close to the melting point of the model (somewhat higher than that of the real solid). We see that the phonons have broadened and a central peak appears to

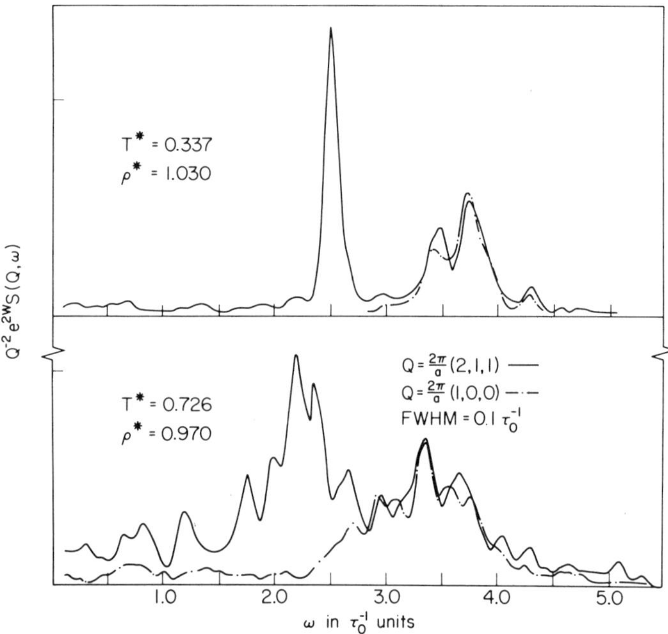

Figure 1. The upper curves show S(Q,ω) data for a 2048 particle Lennard–Jones (12–6) system after 11,000 time steps (of about 10^{-14} secs for argon). The lower curves are for the same Q values but at a higher temperature at lower density after 30,000 time steps.

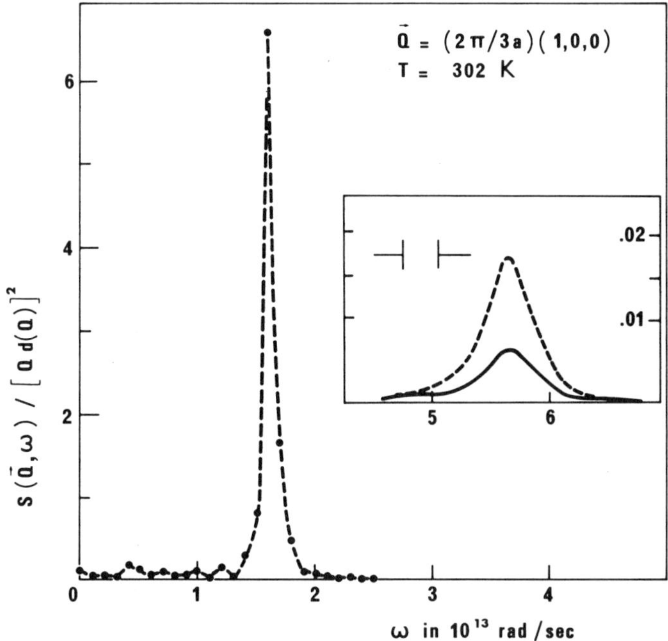

Figure 2. S(Q,ω) *for the smallest accessible wave vector of a system of 216 ions* (Na^+Cl^-) *at 302 K. The dots show the LA response. The inset shows the high frequency LO response on the same relative scale (full line) and the charge density response (dashed line) on an arbitrary relative scale.*

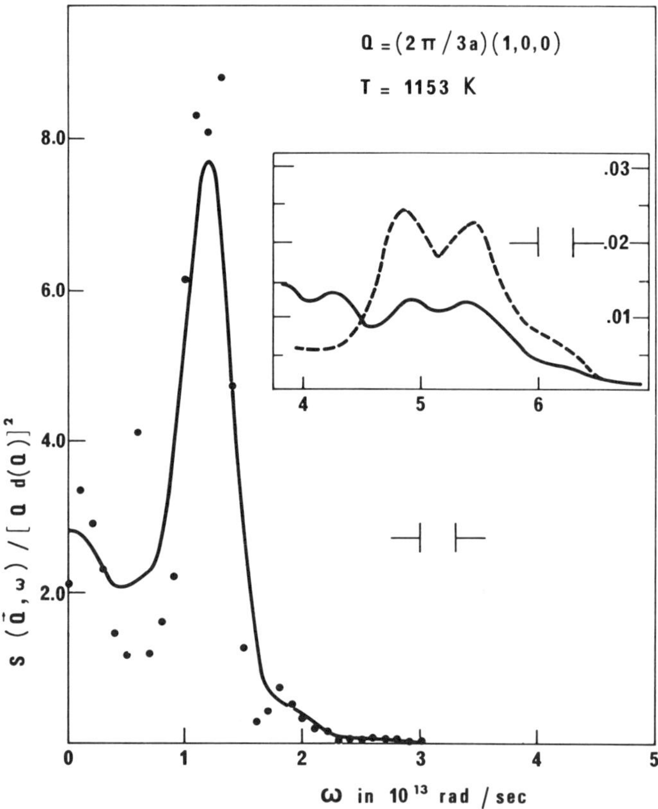

Figure 3. S(Q,ω) for the smallest accessible wave vector of a system of 216 ions (Na^+Cl^-) at 1153 K. The dots show the LA response. The inset shows the high frequency LO response on the same relative scale (full line) and the charge density response (dashed line) on an arbitrary relative scale.

be present also. The LA mode appears to be in reasonable agreement with experimental data where comparison can be made but the LO mode is much too high in frequency due to neglect of polarization effects (22). In subsequent calculations such effects were incorporated via a shell model and a dramatic lowering of the LO peak position was found (23) see Figure 4. Finally, the interference effect between the one phonon peak and the multiphonon background was also studied and several novel effects were predicted. Figure 5 shows the Q dependence of $S(\underset{\sim}{Q},\omega)$ for phonons that would be equivalent in the harmonic approximation or in the one phonon approximation apart from the dull factors of Q^2e^{-2W}. Clearly important contributions arise from the higher order terms in the phonon expansion.

Metals and Alloys

The alkali metals Na and K offer the best prospects for the derivation of effective pair potentials from essentially first principals (24). The method has also been applied to derive potentials for Al. These potentials have been used to calculate the dynamical structure factors $S(\underset{\sim}{Q},\omega)$ for bcc K at temperatures corresponding to approximately one half and nine-tenths melting (15). A typical phonon is shown in Figure 6. The temperature dependence of both the shift and width of this peak agree well with experimental data. Figure 7 illustrates the interference effect between the one phonon peak and the multiphonon background rather nicely. The peaks would be identical in shape in the one phonon approximation. However, the intensity differences on the left hand side of the phonon peak and the shift in peak position are due to the coupling of the one phonon peak to the background. Detailed comparisons with lattice dynamical calculations have been carried out (25).

Some $S(\underset{\sim}{Q},\omega)$ spectra for solid fcc Al close to melting are shown in Figure 8 by the dotted curves there appears to be broadly similar to spectra for fcc Ar suitably scaled (16). Some interesting interference effects have been predicted (16) to be observable in Al. The main interest in these metals has however, centred on the study of defects (26) and their migration by MD methods but such studies go beyond the scope of this article. Finally, the method has been extended to study localised modes of vibration associated with light impurities in alloys. The system $K_{29}Rb_{71}$ has been extensively studied (17) and detailed results will be published in due course. The correlation with available experimental information is excellent. The time evolution of a quenched binary alloy has also been studied using MD (27).

Molecular Crystals

The final class of crystals studied to date are molecular crystals. One feature that is often found is that such crystals

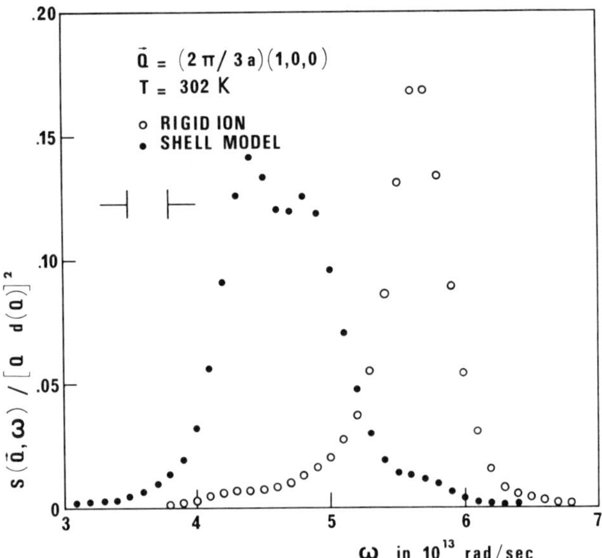

Figure 4. Effect of polarization on the LO mode of smallest wave vector studied. The open circles are for the rigid ion model while the full circles are for the shell model.

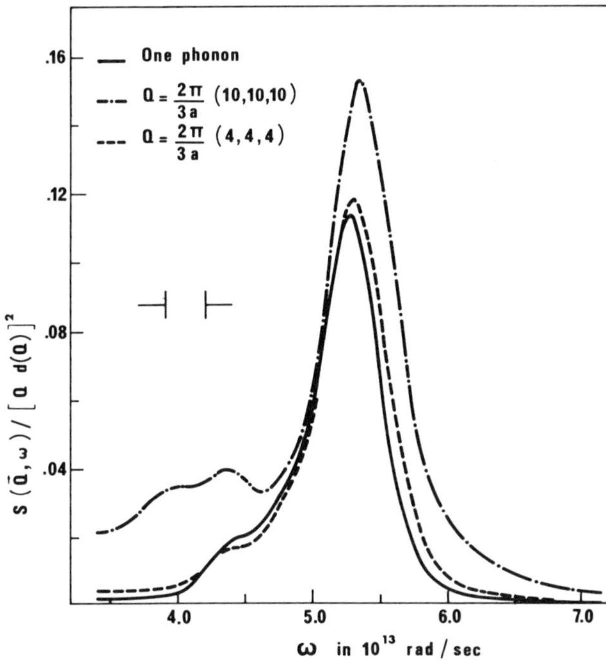

Figure 5. The interference effect in S(Q,ω) for NaCl at 302 K. The one-phonon result is shown by the full curve. The dashed curves is the full S(Q,ω) for the indicated Q value, and shows how the interference effect transfers intensity from the left hand side of the phonon peak to the right hand side. The dashed–dot curve shows a similar effect but for a larger Q value. Here the situation is complicated somewhat by the multiphonon background.

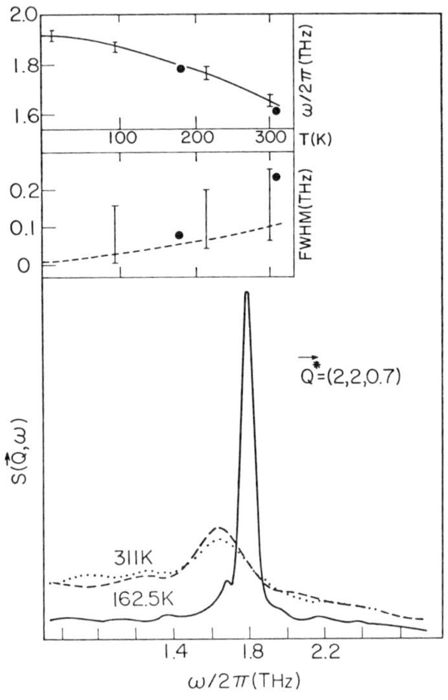

Physical Review

Figure 6. S(Q,ω) for solid K. The solid line corresponds to approximately one half of the melting temperature, the dashed and dotted curves are two independent calculations at about nine-tenths melting. The system consisted of 432 atoms interacting via the potential from Ref. 24 truncated after eight shells of neighbors. The inset shows the temperature variation of the peak position and width compared with experiment (error bars) and lattice dynamical calculations (25).

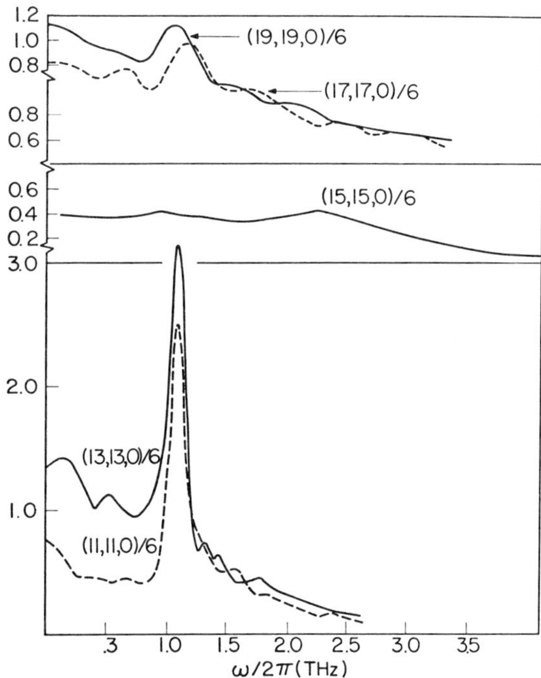

Figure 7. The interference effect in S(Q,ω) of solid K for the indicated Q = Qa/2π values. The lower curves correspond to equivalent phonons either side of the Bragg vector Q* = (2,2,0) while the upper curves refer to similar phonons for either side of the Bragg vector Q* = (3,3,0).*

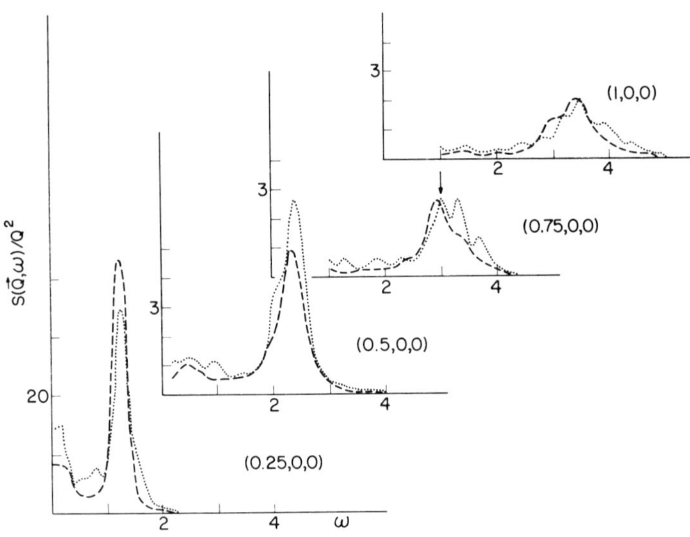

Physical Review

Figure 8. S(Q,ω) data for longitudinal phonons propagating in the <100> direction of fcc Al at 800 K (dotted curves). The unit of frequency is 85 cm^{-1}, 2.55 THz, or 10.6 meV. The dashed curves are similar data for rare gas solids arbitrarily scaled (16).

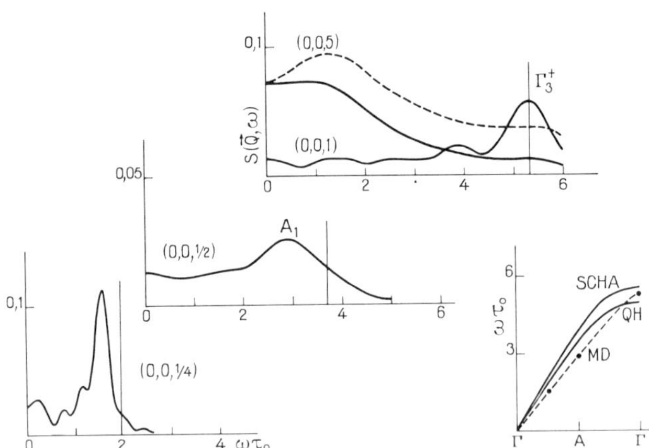

Journal of Chemical Physics

Figure 9. The full curves are the MD S(Q,ω) data for longitudinal phonons propagating along the c-axis of hcp solid β-N_2. Curves labeled QH and SCHA refer to lattice dynamical theories (19). The unit of energy is 16.5 cm^{-1}, 0.50 THz, or 2.05 meV.

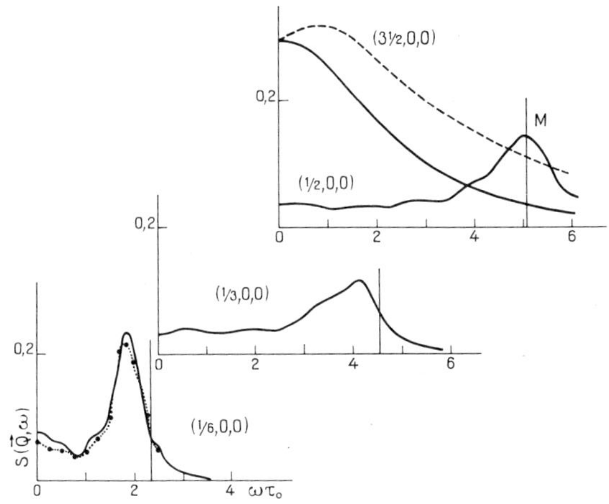

Journal of Chemical Physics

Figure 10. The full curves are the MD S(Q,ω) *data for longitudinal phonons propagating in the basal plane of hcp β-N_2. The dashed dot curve shows the center of mass spectrum. The remnant of the Rayleigh peak is to be noted for the smallest* Q *vector (19).*

undergo structural phase transition in the solid that are associated with differing degrees of rotational freedom of the molecules. The prototype molecular solid is solid Nitrogen. At low temperatures it crystallines in the cubic Pa3 structure (α-phase) with four molecules in the unit cell. When this solid is heated it transforms into the hcp rotational disordered (rotational diffusion or plastic) β-phase. At high pressures and low temperatures yet another phase (γ) occurs that will not concern us here. The nature of the phonon excitations in the α-phase was studied with MD by Weis and Klein (18) using an atom-atom (12-6) potential model. Fair agreement with experiment was obtained at low temperatures but the solid was found to transform into the disordered β-phase too readily since the model molecule did not have a quadrupole moment. The hcp β-phase was also studied (19) and selected $S(\underset{\sim}{Q},\omega)$ spectra are shown in Figures 9 and 10 for longitudinal phonons propagating along the c-axis and in the basal plane respectively. The most striking feature of the data are the enormous changes that occur as $|Q|$ increases and the excitation changes from phonon like motion to predominantly rotational excitation. The model appears to correlate well with experimental (19) data. The most significant finding of the calculations was the importance of rotational-translational coupling in lowering the transverse shear mode frequencies in the disordered β-phase. This probably accounts for the low shear constants observed in all plastic crystals.

Summary

The MD technique has now been used to explore the nature of $S(\underset{\sim}{Q},\omega)$ in a wide variety of crystals and important new insights have been gained. In the future the method will likely be extended to hydrogen bonded solids and other more complicated molecular solids, polymers, superionic conductors (28), liquid crystals, and two-dimensional films (29). Equally novel and important results can be expected there also.

Acknowledgements

It is a pleasure to acknowledge the fruitfull collaborations and friendships I have enjoyed with Jean Pierre Hansen, Giani Jacucci, Ian McDonald and Jean-Jaques Weis that are embodied in this article. The invaluable advice of Dom Levesque and Aneesur Rahman has also been generously given wherever sought after and was much appreciated.

Literature Cited

1. Horton, G.K. and Maradudin, A.A., editors, "Dynamical Properties of Solids" Vols. I and II, N. Holland, Amsterdam (1974,1975).

2. Alder, B.J. and Wainwright, T.E., J. Chem. Phys. (1959), 31, 459.

3. Rahman, A., Phys. Rev. (1964) 136, A405.

4. Levesque, D., Verlet, L. and Kürkijarvi, J., Phys. Rev. (1974), A7, 1690.

5. Rahman, A., Phys. Rev. Lett. (1974) 32, 52.

6. Weis, J.J. and Levesque, D., Phys. Rev. (1976) A13, 450.

7. Hansen, J.P., Pollack, E.L. and McDonald, I.R., Phys. Rev. Lett. (1974) 32, 277.

8. Hansen, J.P. and McDonald, I.R., Phys. Rev. (1974) A11, 2111.

9. Rahman, A. and Stillinger, F.H., Phys. Rev. (1974) A10, 368.

10. Maradudin, A.A. and Fein, A.E., Phys. Rev. (1964) 128, 2589.

11. Hansen, J.P. and Klein, M.L., J. Phys. Lett. (1974) 35, L-29.

12. Dickey, J.M. and Paskin A., Phys. Rev. (1969) 188, 1407.

13. Hansen, J.P. and Klein, M.L., Phys. Rev. (1976) B13, 878.

14. Jacucci, G., Klein, M.L. and McDonald, I.R., J. Phys. Lett. (1975) 36, L-97.

15. Hansen, J.P. and Klein, M.L., Solid. St. Comm. (1976) 20, 771.

16. Jacucci, G. and Klein, M.L., Phys. Rev. (1977) B16, 1322.

17. Jacucci, G., Klein, M.L. and Taylor, R., Solid. St. Comm. (1977) 24, 685.

18. Weis, J.J. and Klein, M.L., J. Chem. Phys. (1975) 63, 2869.

19. Klein, M.L. and Weis, J.J., J. Chem. Phys. (1977) 67, 217.

20. Klein, M.L. and Venables, J.A., editors, "Rare Gas Solids" Vol. I and II, Academic Press, London (1975,1977).

21. Sangster, M.J.L. and Dixon, M., Adv. in Phys. (1976) 25, 247.

22. Cowley, E.R., Jacucci, G., Klein, M.L. and McDonald, I.R., Phys. Rev. (1976) B14, 1758.

23. Jacucci, G., McDonald, I.R. and Rahman, A., Phys. Rev. (1976) A13, 1581.

24. Dagens, L., Rasalt, M. and Taylor, R., Phys. Rev. (1975) B11, 2726.

25. Glyde, H.R., Hansen, J.P. and Klein, M.L., Phys. Rev. (1977) B16, 3476.

26. Da Fano, A. and Jacucci, G., Phys. Rev. Lett. (1977) 39, 950.

27. Sur, A., Lebowitz, J.L., Marro, J. and Kalos, M.H., Phys. Rev. (1977) B15, 3014.

28. Rahman, A., J. Chem. Phys. (1976) 65, 4845.

29. Hanson, F.E., Mandell, M.J. and McTague, J.P., J. Phys. (1977) C-4, 76.

RECEIVED August 15, 1978.

10

Computer Simulations of the Melting and Freezing of Simple Systems Using an Array Processor

G. CHESTER, R. GANN, R. GALLAGHER, and A. GRIMISON

Laboratory of Solid State Physics and Office of Computer Services, Cornell University, Ithaca, NY 14853

The aim of this work is to study, at a microscopic level, precursor phenomenon to freezing and melting. At the present time the studies are being carried out on single phases, crystal and fluid (1,2,3). In the near future we expect to extend some of our work to two phases in equilibria. We hope that these studies will ulitmately lead to an understanding of how the long range spatial order of a crystal is destroyed by melting and how the same order is built up from the chaotic fluid phase. These questions can best be studied initially in the simplest systems. For this reason we have been studying central force classical systems in both two and three dimensions. The differences in dimensions may lead to important insights about the ordering phenomenon.

The properties we plan to calculate by Metropolis (4) Monte Carlo methods are the structure factors S(k) at densities and temperatures near the freezing line, the root mean square displacement of the particles from their lattice sites near melting, the angular correlations between distant pairs of particles in the crystal phase, and near-neighbour correlations in the crystal phase. Several of these properties require very long simulations to achieve good equilibrium values. We have therefore frequently made runs at least one order of magnitude larger than is normal in single phase studies of central potential systems. Determining the size dependence involves a similar amount of computing. In addition, to extend these studies, at least in two dimensions, to two phase simulations requires between 1000 and 10,000 particles and we estimate that runs equivalent to between 3 and 30 hours of CDC 7600 computer time will have to be made. A further consideration is that we plan to complement the Monte Carlo work with molecular dynamics simulations. These again require very long

0-8412-0463-2/78/47-086-111$05.00/0

runs in order to simulate phenomena which occur comparatively slowly in real time. For example the break up of dislocation pairs in two dimensions may only occur on time scales of at least 10^5 conventional time steps of molecular dynamics.

Previous solutions to the problem of such large-scale computations have mainly involved large powerful shared machines (IBM 370, CDC 7600, CDC Star, CRAY 1) or dedicated minicomputers (Prime 400, Harris/4, PDP 11/70, etc.). Each of these solutions has its dedicated advocates, and there are highly favorable features in each case. The large machine offers large memory, disk storage and powerful peripherals, very extensive software, and a fast computation time. The large machines listed above are capable of operating in the range of 2 to 100 megaflops (millions of floating point operations per second). The greatest disadvantage is often the high cost of using such a system, unless it is highly subsidized and then somewhat restricted in access. On a very heavily loaded central system, job turnaround can also be a problem, but the main concern of users of such a system tends to be "How much will it cost?".

The greatest advantage of the dedicated mini-computer is probably the low cost (5), allied with the convenience of control of the resources. This must be offset by the more limited memory and storage available, the limited precision of the hardware in some cases, and the less extensive software libraries available. However, the principal limitation for very large-scale computations stems from the lower computational speed (around 0.5 megaflops or less) so that the users concern is often "How long will it take?".

Recently relatively cheap special purpose "attached processors" have become available, and this paper describes an evaluation which is being carried out at Cornell on the applicability of one such processor to large-scale scientific calculations such as the simulations described earlier. The particular machine is a Floating Point Systems 190-L Array Processor (AP). The AP features include (6,7):

(1) Independent storage for programs, data and constants.
(2) Independent floating point multiplier and adder units.
(3) Two blocks of 32 floating point registers.
(4) Address indexing and counting by independent integer arithmetic unit with 16 integer registers.
(5) Precision enhanced by 38 bit internal floating point format, with hardware rounding.

The very large instruction length and multiple data paths permit parallel operation of all of these elements, so that for example a floating point add, floating point multiply, branch on condition, index increment, memory fetch data memory store, load to and from floating point and integer registers, can be simultaneously executed in one 167 nanosecond cycle, although memory operations cannot be executed in every cycle. The result of this architecture is a processor with a maximum computation rate of around 8 megaflops (6,7), or approximately the raw speed of a CDC 7600 but in the price range of a minicomputer ($60,000 - $140,000). Provided that such a processor can be made to operate near its potential, this clearly provides a very attractive alternative to the maxi- or mini-computer solutions outlined earlier. Since the AP cannot operate independently of a host computer, the first decision that has to be made is the appropriate type of host computer. Earlier models of the AP (Model 120-B) have been attached exclusively to minicomputers. The main concern with such a system is whether the minicomputer is sufficiently powerful to keep the AP operating near its full potential. The average AP configuration has only 32-64K of data memory and 1-2K of program source memory. For large programs and large amounts of data these limits imply a need to load sections (subroutines) of the program into the AP as necessary and to similarly segment and transfer data to and from the AP. Unless the same total calculation can be performed in a lesser clock time than that required by the mini alone, there is no advantage to using the AP except if this frees the mini for other tasks. In other words, assuming the mini is a purchased machine, the key question is still "How long will it take?". A number of installations have been very successfully operating such a configuration in a special-purpose environment for a few years. However, Cornell appears to be the first installation to attach an AP to a large, general purpose host computer and attempt to make it available to a wide range of users.

The Cornell configuration consists of an AP 190-L with 4K of Program Source Memory, 2.5K of Table Memory, and 96K of Main Data Memory, attached to an IBM 370/168 running the Virtual Memory Control Program. Jobs to be run in the AP are prepared using the CMS interactive system and are then spooled to a special virtual machine called the Array Processor Execution Manager (APEMAN) to which the AP is normally attached. APEMAN schedules jobs which require the AP on the basis of a scheduling algorithm including requested priority and length of AP time (jobs can also be placed in hold and released by the user). When a job is selected for initiation, the designated user virtual machine is autologged on, the AP is logically attached to that machine as a normal I/O device, and a designated CMS EXEC (command file) begins to execute in the user's machine. At the end of the run, or when the specified time has elapsed. the AP is detached from the user machine and returned to APEMAN, and the user machine is logged off. The intent of the APEMAN scheduling, when completely implemented, is the normal goal of any operating system or control program; to have the resources of the AP as fully occupied at all times as is practical. In addition, since the Cornell Array Processor is shared among six distinct user groups, APEMAN is used to monitor equitable distribution of the AP for each aggregate group, and will incorporate a priority scheme to facilitate this.

The problem with a large central computer as host is to restrict as much as possible the time spent computing in the host, not because of the elapsed time involved, but because of the cost differential. This paper presents conclusions on the extent and difficulty of reprogramming for the AP, the results of benchmark comparisons for our Monte Carlo calculations, and on the preliminary assessments of the host overhead, as well as strategies to minimize that overhead.

Program Development

The first consideration is program development for the AP. For the AP to approach its potential speed, the programmer has to attempt to keep as many of the parallel operations proceeding simultaneously as possible. A cross-assembler is available to permit program development on the host computer of AP machine code. The optimum strategy appears to be to code interior loops, which involve the bulk of the computation time, first as subroutines executing in the AP. While the subroutine is executing in the AP it is feasible with caution to have simultaneous execution of code in the host. However, with a virtual memory environment like VM, the resident host program will be paged out of memory when inactive and thus incur very little overhead, so we have made no attempt to provide synchronous operation to date. After crucial sections of the program have been optimized for the AP, outer loops can gradually be transferred from host code to AP code. The tradeoff here is that a large amount of programmer time may be spent converting code which accounts for only 5-10% of the overall execution time. In the case of our benchmarks, initially five loops (level 0) were identified which are shown in Table I. These are fairly standard small loops which should be quite similar to those found in most Monte-Carlo simulations. The inner loops were then incorporated in an outer loop (level 1) which generates one attempted move of each of the particles (a pass). For 200 particles, the outer loop is executed 200 times, and each inner loop is also executed 200 times for each cycle through the outer loop. Eventually, another outer loop (level 2) will be transferred to the AP which carries out about 1000-5000 passes of the particles. Previous studies have shown that the level 1 loop encompasses 99% of the total execution time. Recoding the level 0 and level 1 loops, a total of about 75 FORTRAN statements, involved about 100 hours by a programmer who began as a novice (but clearly finished as an expert!). We estimate that this time could be almost halved now.

An important restriction of the AP is the limited amount of program source memory. The "expansion factor" observed for hand-coding was of the order of 4-5 APAL statements for each FORTRAN statement. Since the maximum program source memory available for the AP is 4K, this gives a maximum FORTRAN subroutine length of around 800 statements. It is also clearly very desirable to not

TABLE I

Inner Monte Carlo Loops

Loop	Description

`DO 71 MM=1,NP`
`IF(MM.EQ.M)GOTO 71`
`DX=SIDX2-DABS(DABS(XTRIAL-X(MM))-SIDX2)`
`DY=SIDY2-DABS(DABS(YTRIAL-Y(MM))-SIDY2)`
`R2=DX*DX+DY*DY`
`LSAVE(MM)=R2`
`LL=LSAVE(MM)`
`C=R2-DFLOAT(LL)`
`C1=1.D0-C`
`ESAVE(MM)=C1*RM(LL)+C*RM(LL+1)`
`IND=INDEX(MIN0(MM,M))+MAX0(MM,M)`
`71 DP=DP+ES(IND)-ESAVE(MM)`
	Loop 71 calculates the energy change due to the trial move of a particle and stores the new interparticle distances.

`DO 74 MM=1,NP`
`IF(MM.EQ.M) GOTO 74`
`GN(LSAVE(MM)=GN(LSAVE(MM))+1.D0`
`IND=INDEX(MIN0(MM,M))+MAX0(MM,M)`
`74 ES(IND)=ESAVE(MM)`
	Loop 74 stores the new interparticle energies and saves the interparticle distances for trial moves which lower the energy of the system.

`DO 374 MM=1,NP`
`IF(MM.EQ.M) GOTO 374`
`GN(LSAVE(MM))=GN(LSAVE(MM))+ACC`
`DX=SIDX2-DABS(DABS(X(M)-X(MM))-SIDX2)`
`DY=SIDY2-DABS(DABS(Y(M)-Y(MM))-SIDY2)`
`LL=DX*DX+DY*DY`
`GN(LL)=GN(LL)-ACC1`
`IND=INDEX(MIN0(MM,M))+MAX0(MM,M)`
`374 ES(IND=ESAVE(MM)`
	Loop 374 stores the new interparticle energies and saves the interparticle distances for trial moves which raise the energy of the system.

`DO 274 MM=1,NP`
`IF(M.EQ.MM) GOTO 274`
`GN(LSAVE(MM)=GN(LSAVE(MM)+ACC`
`DX=SIDX2-DABS(DABS(X(M)-X(MM))-SIDX2)`
`DY=SIDY2-DABS(DABS(Y(M)-Y(MM))-SIDY2)`
`LL=DX*DX+DY*DY`
`274 GN(LL)=GN(LL)+ACC1`
	Loop 274 stores the interparticle distances for trial moves which are rejected.

`DO 574 MM=1,NP`
`IF(M.EQ.MM) GOTO 574`
`DX=SIDX2-DABS(DABS(X(M)-X(MM))-SIDX2)`
`DY=SIDY2-DABS(DABS(Y(M)-Y(MM))-SIDY2)`
`LL=DX*DX+DY*DY`
`574 GN(LL)=GN(LL)+1.D0`
	Loop 574 stores the interparticle distances for trial moves which are rejected with an overwhelming probability.

ask a programmer to convert 750 FORTRAN statements which make up the rest of the Monte-Carlo program used here, which only affects 1% of the execution time. Accordingly, Cornell has developed an AP FORTRAN cross-compiler (APTRAN), running on the IBM 370/168. This compiler takes statements written in a natural "computations subset" of FORTRAN, and produces APAL code or optionally AP object code in the appropriate host format. The AP FORTRAN compiler is only intended to remove the burden of programming large amounts of FORTRAN code which do not involve the major portion of the computation time. Even though the compiler does have several levels of optimization, it is still envisaged that hand coding of inner loops will be beneficial. An added advantage of the compiler is that it can reduce the amount of program transfer to and from the AP by chaining calls to AP subroutines. (See later) However, because of the complexity of the compiler it is inevitably a quite large program (more than 10,000 FORTRAN statements). The current "expansion factor" for the compiler is around 10 so that the limited program source memory is an even more serious limitation for compiler-generated code.

Results of the Benchmarks

Benchmarks on the AP were made by the use of a simulator running on the 370/168. Floating Point Systems provides a simulator for program development and debugging, as well as for timing runs. Since the AP is a completely asynchronous processor, such timings should be exact representations of the actual execution times. Simulators are notoriously inefficient, so that it was necessary for long simulations to use a locally modified version of the Floating Point Systems simulator. Because of adaptation to the IBM 370 architecture, this simulator executes 3-4 times faster than the original, and in all runs to date it has produced identical results and timings. The longest run made for these benchmarks simulated a 22 millisecond AP run, and consumed 60 seconds of IBM 370/168 time. The results reported for other computers correspond to actual runs, and in all cases the best optimizing FORTRAN compiler available was used, at full optimization level.

Table II shows the results obtained for some standard benchmarks (5). The values for the CDC 7600 and Harris/4 are from the original literature. In all cases, the timings are in millisecond, obtained from the time required for 1000 iterations of the relevant sequence of instructions. In the floating point benchmark, the AP performed only slightly faster than the IBM 370/168. The following benchmark, involving repeated calls to external arithmetic functions (specifically SQRT, ABS, SIN, ALOGIO, IFIX, ATAN, EXP) is interesting since it illustrates the extremely low overhead incurred by subroutine calls in the AP. The dot product benchmark shows the AP performing very favorably compared to the CDC 7600. This is to be expected, since the particular structure of the algorithm for a dot product is ideally suited to the AP architecture, which permits the simultaneous calculation of the products, and their addition. In fact, it is possible to write the essential structure of a dot product loop in one AP instruction.

Table III shows the results achieved for the actual Monte Carlo inner loops (level 0) detailed in Table I. The results for these inner loops show the AP 190L executing at 0.6 times the speed of the CDC 7600, or at 2.5 times the speed of the IBM 370/168. Table IV compares the AP 190L timings for the level 1 outer loop with those for the CDC 7600, IBM 370/168, and PRIME 400 computers. The execution time, T, of the level 1 loop is a non-linear function of the number of particles considered. It is given by the equation

$$T = AN + BN^2$$

where A and B have been determined for each machine. Timings which were obtained by extrapolation are indicated in parentheses in Table IV. This shows that as the number of particles increases, the speed of the AP 190L for the level 1 loop approaches a limit of 0.63 times the speed of the CDC 7600 or 1.5 times the speed of the IBM 370/168 and 67 times the speed of the PRIME 400.

We believe the lower performance of the AP 190L for the level 1 loop to be attributable to a number of factors. Whereas in suitable applications the AP 190L has provided speeds above that of a CDC 7600 (7), in general the limiting factor in loop speed is data memory access time. In this initial version of the Monte Carlo program, a great deal of storing and fetching variables and base addresses was performed between the level 0 loops. No attempt has yet been made to optimize the portions of the level 1 loop around the inner loops. An additional consideration was the fact that the existing Monte Carlo FORTRAN program which was converted for the AP was not at all optimally structured for the AP. It is the case that the results for the AP were obtained from hand coded routines, while those for the CDC 7600 and IBM 370/168 computers were obtained from the FTN4 and H Extended optimizing compilers respectively. However in our experience these optimizing compilers have proved extremely efficient for the small loops involved here, while there are extensive opportunities for further hand optimization of the AP code. For these reasons we believe the factor of 0.42 times the CDC 7600 speed to be a lower limit of the capability of the AP for this application.

TABLE II

General Benchmarks

	CDC 7600	Harris/4	IBM 370/168	AP 190L
Integer Arith.	1.8	41	4.3	NA
Floating Point	0.21	4.8	0.60	0.55
External Funct.	0.32	4.1	0.97	0.26
Dot Product	0.70	30	1.50	0.84

These benchmarks are taken from reference 5.

TABLE III

Monte Carlo Benchmarks

LOOP #	CDC 7600 TIME (MS)	IBM/370 TIME (MS)	AP TIME (MS)
71	.597	2.2	.885
74	.316	1.2	.634
374	.583	1.9	.885
574	.487	1.4	.300
274	.448	1.6	.601
TOTAL	2.431	8.3	3.305

All times are in milliseconds and measure the time needed to execute the stated loop 100 times.

TABLE IV

Comparative Times for Execution of

N-Particle Monte Carlo Program

(milliseconds)

#Particles	PRIME 400	IBM 370/168	CDC 7600	AP 190L
16	(416.0)	6.8	(2.0)	4.9
36	1650.0	30.2	(9.2)	21.9
64	4710.0	98.2	28.0	(66.3)
100	($1.09\ 10^3$)	243.0	67.2	(158.5)
1000	($1.00\ 10^6$)	($2.32\ 10^4$)	($6.55\ 10^3$)	($1.53\ 10^4$)
10000	($9.92\ 10^7$)	($2.31\ 10^6$)	($6.53\ 10^5$)	($1.52\ 10^6$)

Host Overhead

The full cost-recovery rate for the AP-190L at Cornell, based on a three-year amortization schedule, is $40/hour. Since the previous benchmarks have shown that the AP can run at from 1.5-2.5 times the speed of the IBM 370/168, this yields a cost-comparison of about 80:1 in the Cornell environment. It is important to note that this ratio uses the prime-time 370 rates, and can be reduced by a factor of about 0.4 for overnight turnaround. The comparison may also be substantially less favorable to the AP for a partially-subsidized host computer. However, this is still a dramatic demonstration of the potential cost advantage of carrying out large-scale scientific calculations on such an attached processor.

The magnitude of the cost-differential makes the problem of host overhead even more crucial, since the cost of a run with 90% of the computation in the AP and 10% of the computation in the host could still be totally dominated by the host charges. Host overhead is incurred in a number of ways:

(i) Job submission.
(ii) System overhead in making the AP available to the particular user.
(iii) Maintenance of a main program resident in the host (AP programs can only be executed as subroutines called from the host).
(iv) CALL to the AP subroutine from the main program.
(v) Initialization of the AP.
(vi) Transfer of program source microcode to the AP for every subroutine call, including the updating of location tables, etc.
(vii) Transfer of data to and from the AP.
(viii) Release the AP for a subsequent CALL.
(ix) RETURN from the AP subroutine.
(x) Release the AP for other users.

Table V details the typical host overhead involved in the execution of an AP program at Cornell. Due to the characteristics of the Cornell Virtual Machine environment mentioned earlier, there is no appreciable overhead (less than 1 msec) to maintain a main program resident in the host, since the program is paged out of real memory when inactive. It is important to note that the majority of the contributions to the total overhead are fixed quantities. Exceptions which are under the control of the programmer include the number of data transfers (APPUTs and APGETs), and the number of subroutine calls. One major objective of the Cornell APTRAN compiler and its associated linkage editor was to minimize data transfers by transferring data in COMMON blocks in one chained operation (without explicit transfer being required by the programmer, as is otherwise the case). In addition, APTRAN chains AP subroutine calls together, effectively replacing the CALL and RETURN, program transfer, and data transfer (in some

TABLE V

Typical Values for Host Overhead (IBM 370/168)

Operation	Approx. Overhead (msec.)	Comments
APM SUBMIT	10	Job submission
AP ATTACH	40	Attach AP to user machine
Subroutine CALL	0.5	
APINIT	58	Initialize the AP
APPUT	1	Data Transfer to AP (each)
APGET	1	Data Transfer from AP (each)
Transfer AP code	10	Put routine's micro-code in AP (up to 4K instructions)
APRLSE	1	Release AP for subsequent CALL
Subroutine RETURN	0.5	
APDETACH	10	Relinquish AP from user machine
TOTAL	132 msec.	

cases) by an AP "JUMP SUBROUTINE" instruction. This instruction was shown earlier to incur an extremely low overhead (167 nanosecond worst case).

The Table shows that a minimum overhead of about 150 msec. of 370/168 cpu time must be involved in any AP computation, no matter how trivial. The cost ratio of 80:1 leads to a rule-of-thumb estimate that the full cost benefits of the AP will not be approached unless a subroutine can be created which:

(i) Contains less than 300-400 FORTRAN statements (because of the limited Program Source Memory) if using a compiler. This number might be doubled for hand-coding.

(ii) Operates on less than 64K floating point numbers as data at one time (96K for hand-coding).

(iii) Computes for at least 1-2 minutes in the AP (3-5 minutes on an IBM 370/168 or 30-90 seconds on a CDC 6700) between subroutine CALL and RETURN. (This does not imply that intermediate data transfers are not made to and from the AP resident subroutine).

Point (iii) above may need some clarification; the largest part of the overhead per subroutine call is involved in initializing the AP and transferring the microcode for the subroutine. Once the subroutine is operating in the AP, it can make data transfers to and from the host at an overhead of about 1 msec. per transfer (independent of the amount of data transferred up to the capacity of the AP). Thus careful programming can ensure that the same subroutine remains resident in the AP, but that this subroutine operates on sucessive batches of data; it is the return from the subroutine to the host and the corresponding call which must be minimized.

For applications which fit the above constraints the cost advantages of the AP can be outstanding. For example, in the Monte-Carlo calculations described in this paper the overall program strategy is still being refined to minimize subroutine calls, but it appears that a goal of over 99% of the computation time in the AP and less than 1% 370/168 overhead can definitely be realized. This in turn implies an overall cost in the range of $50/AP hour, even using the prime time rates on the IBM 370/168.

Accuracy of the Calculations

One concern with potential users of the AP 190L (or the equivalent AP 120B) for scientific work is the available word length. No hardware double precision capability is available on the AP 190L, so that all floating point operations are carried out in the 38 bit internal floating point format, with a 10 bit binary exponent, a 28-bit mantissa, and a hardware convergent rounding algorithm in the Floating Adder and Floating Multiplier. Floating Point Systems (6) state that this provides a precision of 8.1 decimal digits, as compared to 7.2 decimal digits for IBM 370 single precision, 16.8 decimal digits for IBM 370 double precision, and approximately 14 decimal digits for the CDC 7600 in single precision.

Table VI shows the results of some accuracy comparisons which we have made. The dot product results correspond to a dot product of two vectors of 1000 elements, with random values between 0 and 1. This shows that the AP 190L result gives the first 7 figures of the results from the IBM 370 double precision calculation. The other entries in Table VI show results from the Monte Carlo calculations.

TABLE VI

Effect of AP Word Length on Precision

	IBM REAL*8	AP 190L
Dot Product	0.255178129E03	0.255178122E03
Total Energy (16)	-30.7006546	-30.7006483
Total Energy (36)	-61.6051765	-61.6052179

The total energy is the running average energy computed for one pass through the system of 16 or 36 particles. Maintaining the accuracy of the running average energy is crucial in Monte Carlo calculations. For large systems of several hundred particles it may be necessary to recompute the running average in the course of a pass, since this quantity is the most sensitive to accumulated errors. This shows that the AP 190L result gives the first 7 figures of the IBM 370 double precision result for 16 particles, and the first 6 figures for 36 particles.

After testing the AP random-number generator routine available in the Floating Point Systems Math Library, a random-number generator for a 28-bit mantissa architecture was written in FORTRAN and compiled on the APTRAN compiler.

Conclusions

In our experiments the Floating Point Systems 190L Array Processor proved capable of speeds between 0.43 and 0.66 the speed of a CDC 7600. The broad conclusion is that the AP 190L appears to be between 10 and 100 times more cost-effective than the other systems we have discussed for the type of Monte Carlo simulations we report. In fact the cost-effectiveness of the AP 190L is such that there is little economic incentive to extensive program optimization, though it may be esthetically satisfying. While it will be necessary later to include the additional costs involved when actually using the combined IBM 370/AP 190L system, we do not estimate that these costs will materially alter our conclusion. It appears also that the word length of the AP 190L is sufficient to provide adequate precision for our purposes.

Acknowledgements

This work was supported by the National Science Foundation Grant No. DMR-74-23494 and also through the Materials Science Center, Cornell University, through the National Science Foundation Grant No. DMR-72-03029. The Office of Computer Services provided IBM 370/168 computer time. We would also like to thank Mr. J. Tobochnik for help in running the benchmarks on the PRIME computer.

Literature Cited

1. Ceperley, D., Chester, G.V. and Kalos, M.H., Phys. Rev., (1977), 16, 3081.
2. Ceperley, D., Chester, G.V. and Kalos, M.H., Phys. Rev., (1978), 17, 1070.
3. Gann, R., Chakravarty, S. and Chester, G.V., Phys. Rev., (1978), to be published.
4. Metropolis, N., Rosenbluth, A.W., Rosenbluth, M.N., Teller, A.H. and Teller, E., J. Chem. Phys. (1953), 21, 1087.
5. Schaefer, H.F. and Miller, W.H., Computers and Chemistry, (1971), 1, 85.
6. Floating Point Systems "AP 120-B Processor Handbook", 7259-02. Floating Point Systems, Beaverton, Oregon 1976.
7. Bucy, R.S. and Senne, K.D., "Nonlinear Filtering Algorithms for Parallel and Pipe Line Machines". Proceedings of the Meeting on Parallel Mathematics - Parallel Computers, Munich March 1976, North Holland Press, Amsterdam.

RECEIVED August 15, 1978.

11

Simulating the Dynamic and Equilibrium Properties of a Multichain Polymer System

DAVID E. KRANBUEHL and BRUCE SCHARDT

Department of Chemistry, College of William and Mary, Williamsburg, VA 23185

The dynamic behavior of polymer chains continues to be the subject of numerous molecular model studies in the polymer literature. The basis for most of this work is the bead-and-spring model of Rouse (1). In the Rouse model, a polymer chain is represented by a large number of connected segments. Each segment is statistically equivalent to a sufficient number of monomer units such that the overall motion of the polymer may be idealized by the movement of these freely jointed links. This model with its associated modifications for hydrodynamic interaction and excluded volume has achieved considerable success in describing dielectric, neutron scattering, nuclear magnetic and visco-elastic-mechanical properties of polymer systems (2-10). A major limitation of these analytical calculations is the difficulty of introducing specific molecular interactions without approximating or pre-averaging parameters of physical interest.

Another approach used to investigate dynamic behavior is to simulate Brownian motion using Monte Carlo techniques on a lattice-model polymer chain with the aid of a high speed digital computer (11, 12, 13, 14, 15). In our lattice model the configuration of a polymer chain N-1 units long is represented by a string of N connected points on a 3-dimensional lattice. The points referred to as beads lie on the vertices of the lattice. The links between the beads representing a large number of monomer units are each of unit length. Brownian motion of the chain resulting from random collisions with solvent molecules is simulated by choosing beads at random and then moving them to a new position on the lattice according to fixed bead movement rules which maintain chain connectivity and a bead segment distance of one lattice unit. As this process, which is called a bead cycle, is repeated again and again, the chain moves from one conformation to another, eventually (in principle) taking on all the conformations in its equilibrium ensemble. Periodic sampling of the positions of the beads and calculation of quantities of experimental interest give estimates of equilibrium ensemble averages. Relaxation behavior in the equilibrium ensemble is studied by storing these

0-8412-0463-2/78/47-086-125$05.00/0

quantities along with the elapsed time in bead cycles and then calculating time-correlation functions for the chain properties of interest. Bead cycle time may be related to real time by measuring the translational diffusion constant of the polymer chain.

We regard these dynamic Monte Carlo lattice chain simulations as computer "experiments" which complement analytical calculations. A lattice chain model has been widely used in both analytical and numerical calculations to describe equilibrium properties of polymers (16). Numerical calculations have been particularly helpful in treating equilibrium properties which are difficult to handle analytically (17, 18, 19). Similarly, the Monte Carlo lattice chain model is often more advantageous than the analytical model of Rouse in the treatment of dynamic behavior. The dynamic behavior of the Monte Carlo lattice model is similar to the Rouse model (20, 21, 22). However, the Monte Carlo lattice model facilitates the introduction of specific types of interaction such as excluded volume and the evaluation of quantities which are not readily amenable to analytical treatment (11,23,24). Thus, the Monte Carlo lattice chain model should provide a useful approach for studying the dynamic and equilibrium properties of polymers as a function of concentration, currently an important and particularly difficult problem in polymer chemistry.

Model

Details of the dynamical model for a single linear polymer chain on a simple cubic lattice have been previously described by Verdier (25). The model has been extended so that the dynamic properties of multi-chain polymer systems can be simulated. In the model, a polymer chain is represented by a string of N connected points on a cubic lattice. A collection of n polymer chains each of length N are placed in a hypothetical cubic space of volume L^3 with penetrable walls. When a polymer bead moves through one of the walls, it is relocated a distance L so that it reappears on the opposite side of the box. Thus, the number of beads in the volume L^3 remains constant. This procedure introduces an artificial periodicity of period L into the system as shown in Figure 1.

Brownian motion of the chains is simulated by choosing beads in the box at random. For non-end beads, the chosen beads position r_j is moved to a new position r_j' according to

$$r_j' = r_{j-1} + r_{j+1} - r_j,$$

where r_j is a vector from an arbitrary position to the jth bead. If an end bead is picked, it moves to one of four possible new positions r_{end}' such that

$$(r_{end}' - r_{next}) \cdot (r_{end} - r_{next}) = 0.$$

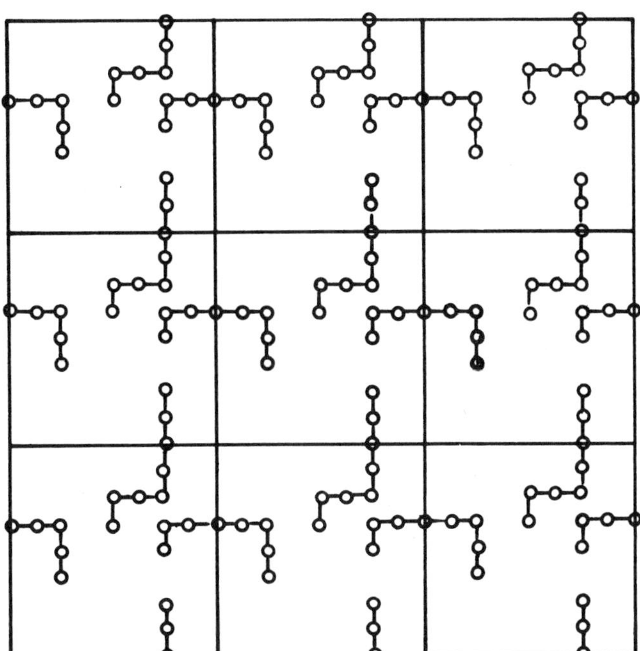

Figure 1. Two eight-bead chains

This process, which may be thought of as corresponding to 1/nN units of time, is referred to as a bead cycle. The effect of chain entanglements and excluded volume is accounted for by not allowing the bead to move to its new site if this site is already occupied by another bead from one of the chains.

In this study relaxation from equilibrium positions was examined. Five to ten computer runs each were made on multi-chain systems for N=10 and N=20 at varying concentrations. The final configuration of the chains in each run served as the initial configuration for the next run. The total time interval of each run consisted of 100 to 1000 equal subintervals. Each subinterval or frame lasted between 1.0 to 1.5 nN^3 bead cycles, a time which is comparable to the relaxation for the vector end-to-end length. Since vector end-to-end length appears to relax on the time scale of the slowest motions of a chain, the number of frames represents a crude estimate of the number of independent samples represented in each run.

At the start of each frame, a time denoted as t_o, the instantaneous value $\ell(t_o)$ of the vector end-to-end for each chain in the box and its square $\ell^2(t_o)$ were sampled and saved. At each of a series of later time intervals t_o+t, ℓ and ℓ^2 were again sampled and multiplied by the corresponding values at t_o to give sampled values of the products $\ell(t_o) \cdot \ell(t_o+t)$ for each chain. These products along with sampled values of ℓ^2 obtained at the same time were added into running sums. At the end of each run, the running sums were converted to averages and used to give estimates of the ensemble averages $\langle \ell(t_o) \cdot \ell(t_o+t) \rangle$, $\langle \ell^2 \rangle$ and $\langle \ell^4 \rangle$. The estimates from all runs and all the chains in the box were combined. The autocorrelation function in ℓ was calculated as

$$\rho(\ell,\ell,t) = \langle \ell(0) \cdot \ell(t) \rangle / \langle \ell^2 \rangle .$$

Results and Discussion

Table I gives the values of ℓ^2, α^2 and ℓ^4 for each simulation. The estimate of the uncertainty is the standard deviation of the average over all the runs. The values of the expansion factor squared are computed as $\alpha^2 = \langle \ell^2 \rangle/(N-1)$, where N is the number of beads per chain. The bead density or polymer volume fraction φ is the total number of beads divided by the number of lattice sites. All values of $\langle \ell^2 \rangle$ for the multiple 10 and 20 bead simulations agree to within one standard deviation of previously published equilibrium data.

De Vos and Bellemans (26) examined the chain length dependence of $\langle \ell^2 \rangle$ for N=7, 11, 21 and 31 bead chains as a function of the polymer volume fraction φ. More recently, Wall, Chin and Mandel (27) have examined $\langle \ell^2 \rangle$ for chains up to 105 beads. The dependence of $\langle \ell^2 \rangle$ on φ has been extrapolated to the bulk phase

TABLE I: EQUILIBRIUM DATA

#RUNS	FRAMES/RUN	FRAME LENGTH nN^3 BEAD CYCLES	BOX SIZE	POLYMER VOLUME FRACTION φ	$\langle \ell^2 \rangle$	α^2	$\langle \ell^4 \rangle$
MULTIPLE 10 BEAD CHAINS							
10	1000	1.0	8x8x8	0.0195	14.8± .1	1.643	297± 8
10	500	1.0	6x6x6	0.1852	14.4 .2	1.597	282 7
10	400	1.0	6x6x6	0.4167	13.9 .3	1.543	267 9
10	400	1.25	6x6x6	0.6019	13.5 .3	1.506	257 11
5	400	1.5	7x7x7	0.6122	13.5 .3	1.496	255 10
MULTIPLE 20 BEAD CHAINS							
10	200	1.25	8x8x8	0.0391	35.7± .9	1.882	1800± 70
5	100	1.25	8x8x8	0.1953	34.1 1.4	1.793	1660 140
5	100	1.25	7x7x7	0.4082	31.6 1.5	1.665	1440 130
5	100	1.25	7x7x7	0.5248	30.9 1.6	1.613	1380 130

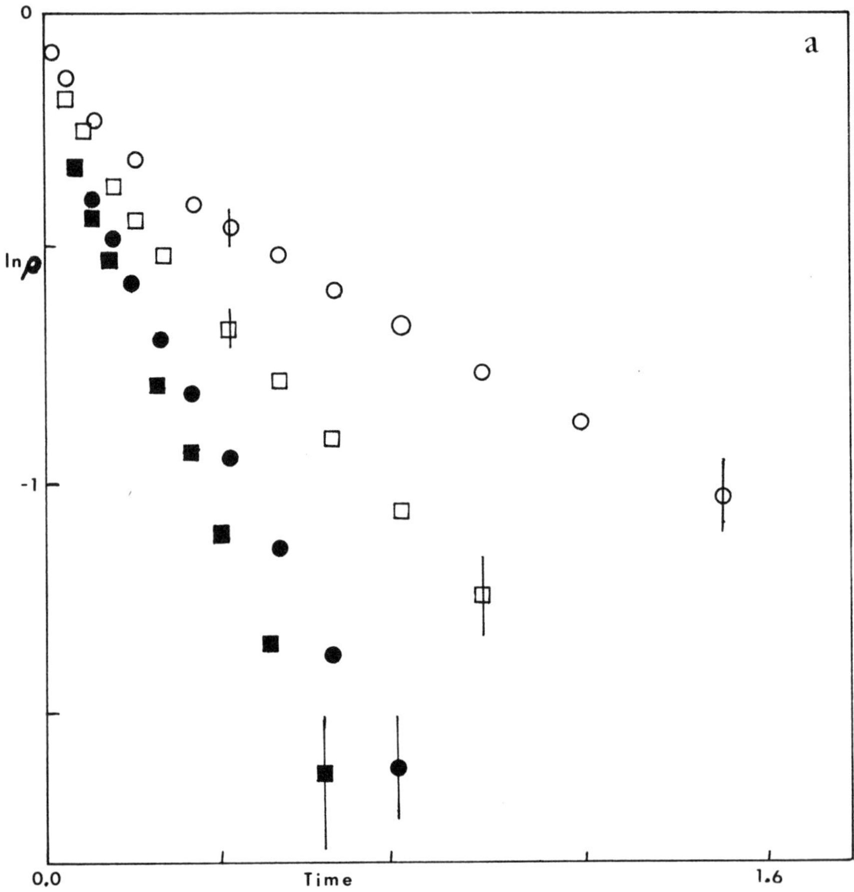

Figure 2. ln(ρ) vs. time in units of nN³ *bead cycles. (a) Multiple 10-bead chains. Polymer volume fraction φ. (■) 0.0195, (●) 0.1852, (□) 0.4167, (○) 0.6122. (b) Multiple 20-bead chains. Polymer volume fraction φ. (■) 0.0391, (●) 0.1953, (□) 0.4082, (○) 0.5248.*

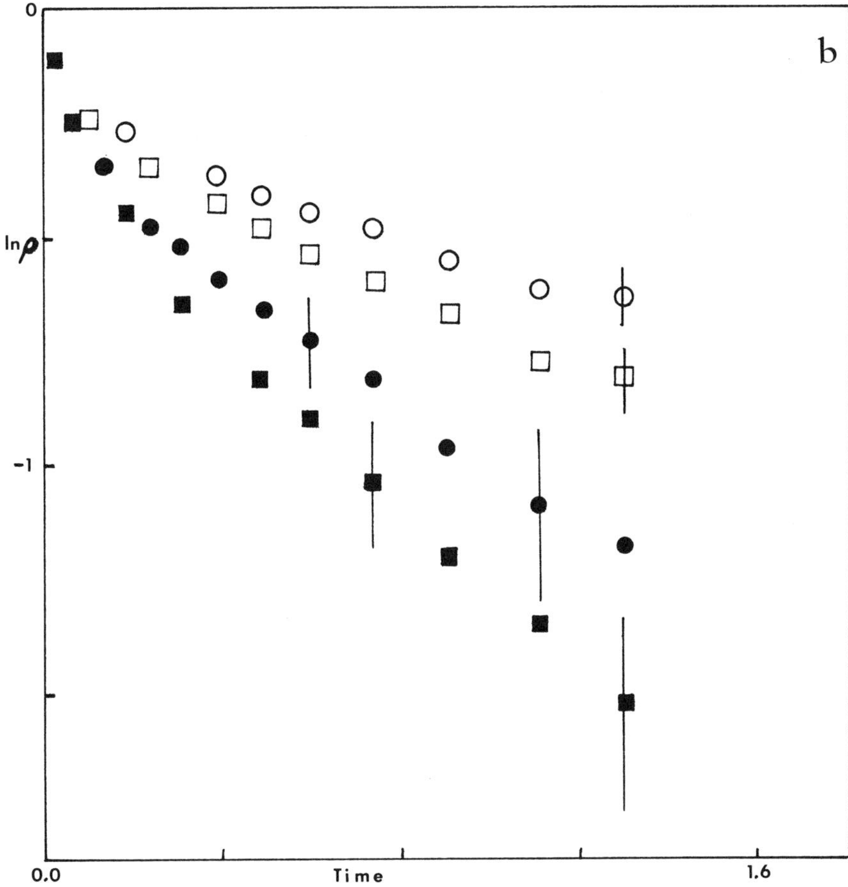
b
0
ln ρ
-1
0.0
Time
1.6

limit $\varphi = 1$. Bellemans' data suggests that $\langle \ell^2 \rangle \sim (N-1)^{1.07}$ when $\varphi = 1$. This should be compared with $\langle \ell^2 \rangle \sim (N-1)^{1.20}$ for a single chain with excluded volume and $\langle \ell^2 \rangle \sim N-1$ for a Gaussian chain. Analytical and experimental data support the conclusion that the equilibrium value of $\langle \ell^2 \rangle$ approaches gaussian behavior in the bulk phase (28, 29). Flory first proposed this result and argued that in the bulk phase the polymer chain has nothing to gain by expanding since a decrease in interaction with itself is compensated for by increased interference with its neighbors.

The autocorrelation functions $\rho(\ell,\ell,t)$ of vector end-to-end length for 10 and 20 bead chain systems at various bead densities are shown in Figure 2. The plot shows $\ln\rho(\ell,\ell,t)$ versus time in nN^3 bead cycles. N is the number of beads in the chain and n is the number of chains on the lattice. Although the autocorrelation functions $\rho(\ell,\ell,t)$ as shown in Figure 2 are not simple exponentials, we arbitrarily define a relaxation time $\tau_{1/e}$ as the time interval in which $\rho(\ell,\ell,t)$ drops to 1/e, i.e. $\rho(\ell,\ell,\tau_{1/e}) = 1/e$. In order to characterize the relaxation behavior of vector end-to-end length, we have extracted two other time parameters from each autocorrelation function in Figure 2. At long times, the semilog plots appear to approach linearity. We estimate the longest relaxation time τ_s and its contribution A to the relaxation from the limiting slope of $\ln\rho(\ell,\ell,t)$ versus time. The limiting slope relaxation time τ_s should be dependent on the longer range cooperative motion of the chains. Thus, τ_s is the parameter of greatest interest in examining the effect of concentration on the dynamic behavior of the polymer chain.

Table II lists the values of A and τ_s and $\tau_{1/e}$ for the 10 and 20 bead multiple chain simulations in units of nN^3 bead cycles. For chains without excluded volume, τ_s in units of nN^3 bead cycles is approximately constant. As shown in Table II, we observe first that τ_s increases as the polymer volume fraction increases. For the 20 bead chains, the contribution of the longest relaxation time τ_s to $\rho(\ell,\ell,t)$ as measured by A increases back toward the no excluded volume result as the polymer volume fraction increases. No trend in A for the 10 bead chains is observed.

Figure 3 shows a plot of τ_s versus the polymer volume fraction φ. The non-linear behavior of τ_s versus φ reflects the non-uniform density of polymer beads on the lattice at low concentrations. At low values of φ, the beads are clustered in n units, the number of chains, each of N beads. This makes the local bead density much greater than the lattice average. As the polymer volume fraction φ increases, the local bead density does not rise as fast. As φ increases further, a point is reached when increasing the bead density begins to have a pronounced effect on the number of interchain excluded volume conflicts. At this point, τ_s rises sharply with an increase in φ.

Free volume has been a particularly useful concept to describe dynamic properties of polymer systems. Assuming that the

TABLE II: RELAXATION TIMES FOR VECTOR END-TO-END LENGTH

CHAIN LENGTH	# OF CHAINS	POLYMER VOLUME FRACTION φ	A	τ_s*	$\tau^*_{1/e}$	$\frac{\varphi}{1-\varphi}$	$\ln \tau_s$
10	1	0.0195	0.82	0.44	0.35	0.020	-0.82
10	4	0.185	0.80	0.56	0.44	0.228	-0.58
10	9	0.417	0.79	0.96	0.73	0.717	-0.04
10	13	0.602	0.79	1.76	1.36	1.51	0.57
10	21	0.612	0.79	1.90	1.45	1.58	0.64
20	1	0.039	0.69	1.13	0.71	0.041	0.12
20	5	0.195	0.73	1.45	1.00	0.242	0.37
20	7	0.408	0.79	2.06	1.57	0.649	0.72
20	9	0.525	0.80	2.77	2.15	1.10	1.02

*units of nN^3 bead cycles
Uncertainty in τ 20%
Uncertainty in A ± .08

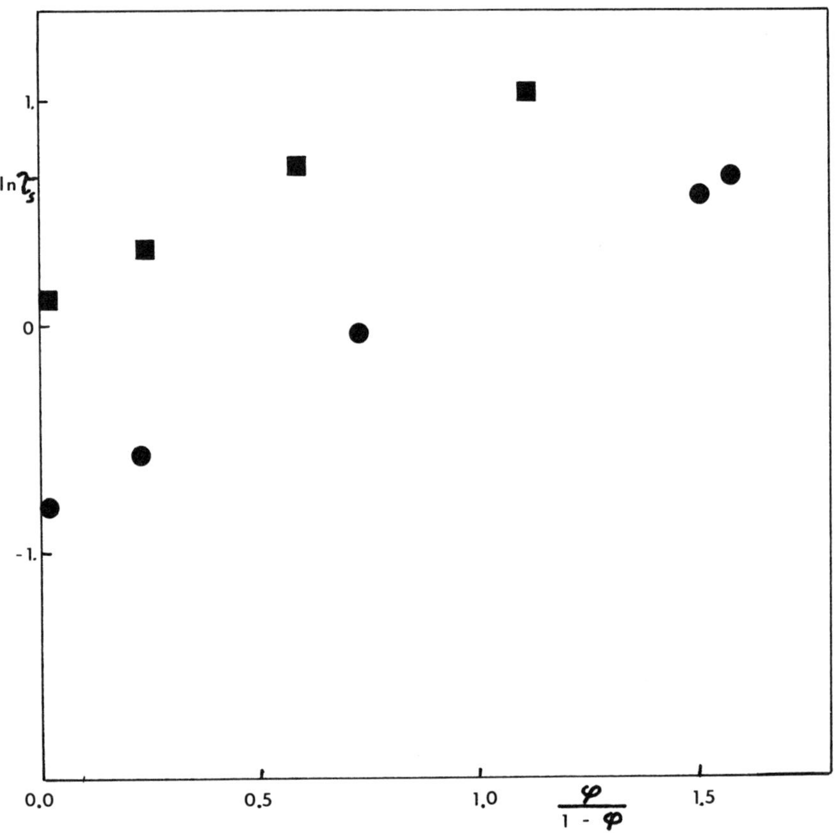

Figure 3. *$ln(\tau_s)$ vs. $\varphi/(1-\varphi)$. (●) N = 10 chains, (■) N = 20 chains.*

dependence of τ_s and the viscosity are similarly dependent on long time cooperative motions of the polymer chains, we may write as suggested by Doolittle (30) and Williams, Landel and Ferry (8, 31)

$$\tau_s = A \exp (BV_o/V_f) = A \exp (B\varphi/(1-\varphi))$$

where A and B are constants, V_o is the volume occupied by the polymer molecule, and V_f is the average free volume per molecule. Table II lists values of ln (τ_s) and $\varphi/(1-\varphi)$. Figure 3 shows that a plot of ln(τ_s) versus $\varphi/(1-\varphi)$ for the 10 and 20 bead multiple chain systems is linear as predicted by the theory. Thus, over the range of polymer volume fractions studied, the free volume theory provides a good description of the dependence of the relaxation time on polymer volume fraction.

Only two chain lengths have been examined. It is appropriate to make only tentative observations about the N dependence of τ_s as the density φ of the chains increases. Earlier results on single chain systems with excluded volume interactions showed that $\tau_s \sim N^4$. This result may be compared with the Rouse, no excluded volume result for a single chain that $\tau_s \sim N^3$. Comparing the 10 and 20 bead chain systems with excluded volume, the N dependence of τ_s does not appear to be affected by an increase in φ in the concentration range studied. Further work at higher concentrations and for longer chain lengths N is in progress.

Partial financial support from the Petroleum Research Corporation administered by the American Chemical Society is gratefully acknowledged.

Literature Cited

1. Rouse, P. E., Jr., J. Chem. Phys. (1953) 21, 1272.
2. Zimm, B. H., Jr., J. Chem. Phys. (1956) 24, 269.
3. Pyun, C. W. and Fixman, M., J. Chem. Phys. (1965) 42, 3838.
4. Tshoegl, N. W., J. Chem. Phys. (1964) 40, 473.
5. De Gennes, P. G., Macromolecules (1977) 9, 587 and 594.
6. Stockmayer, W. H., "Fluides Moleculaires" (Gordon Breach, New York, 1976).
7. Gupta, S., Shah, V. and Forsman, W., Macromolecules (1974) 7, 948.
8. Ferry, J. D., "Viscoelastic Properties of Polymers" (Wiley, New York, 1970).
9. McCrum, N. G., Read, B. F. and Williams, G., "Anelastic and Dielectric Effects in Polymeric Solids" (Wiley, New York, 1967).
10. Stockmayer, W. H., Pure Appl. Chem. (1967) 15, 539.
11. Verdier, P. H. and Stockmayer, W. H., J. Chem. Phys. (1962) 36, 227.
12. Verdier, P. H., J. Chem. Phys. (1966) 45, 2122.
13. Valeur, B., Jarry, J., Geny, F. and Monnerie, L., J. Polym.

Sci., Poly. Phys. (1975) 13, 667 and 675.
14. Taran, Yu A., Dokl. AN SSR (1970) 191, 1330.
15. Kranbuehl, D. E. and Verdier, P. H., J. Chem. Phys. (1972) 56, 3145.
16. Yamakawa, H., "Modern Theory of Polymer Solutions" (Harper & Row, New York, 1971).
17. Wall, F., Chin, J. and Mandel, F., Macromolecules (1977) 7, 3143.
18. Rubin, R. and Weiss, G., Macromolecules (1977) 10, 332.
19. Gobush, W., Solc, K. and Stockmayer, W. H., J. Chem. Phys. (1974) 60, 12.
20. Orwoll, R. A. and Stockmayer, W. H., Advan. Chem. Phys. (1969) 15, 305.
21. Verdier, P. H., J. Chem. Phys. (1970) 52, 5512.
22. Iwata, K. and Kurata, M., J. Chem. Phys. (1969) 50, 4008.
23. Kranbuehl, D., Verdier, P. and Spencer, J.,J. Chem. Phys. (1973) 59, 3861.
24. Kranbuehl, D. and Verdier, P., J. Chem. Phys. (1977) 67, 361.
25. Verdier, P., J. Comp. Phys. (1969) 4, 204.
26. De Vos, E. and Bellemans, A., Macromolecules (1975) 8, 651.
27. Wall, F., Chin, J. and Mandel, F., J. Chem. Phys. (1977) 66, 3143.
28. Fixman, M. and Peterson, J., J. Am. Chem. Soc. (1964) 86, 3524.
29. Colton, J. et al, Macromolecules (1974) 7, 863.
30. Doolittle, A., J. Appl. Phys. (1951) 22, 1471; ibid. (1952) 23, 236.
31. Williams, M., Landel, R. and Ferry, J., J. Am. Chem. Soc. (1955) 77, 3701.

RECEIVED August 15, 1978.

12

Application of Conformational Energy Calculations to Defect Properties in Polymer Crystals

RICHARD H. BOYD

Department of Materials Science and Engineering, and Department of Chemical Engineering, University of Utah, Salt Lake City, UT 84112

The conformational energy or "molecular mechanics" method has proven to be highly successful in predicting structures, stabilities and other properties of isolated organic molecules (1). The model basically assumes that molecular energetics (conformational energy) can be represented by empirical transferable energy functions that describe bond stretching, bending and twisting and also non-bonded interactions. Molecular geometries are those of local minima in the total conformational energy. Thus the problem of structure calculation becomes that of seeking local minima in a function with variables equal to the number of internal degrees of freedom of the molecule. Although this problem is far from trivial, algorithms have been developed which allow the more or less routine use of the computer in accomplishing this. These include adaptations of the not so efficient steepest descent method (2) and faster Newton-Raphson (3,4,5), Fletcher-Reeves (6) and relaxation (7) methods. Furthermore, the energetics of transitions between different conformations can often be computed by constraining certain internal coordinates to a succession of fixed values which "track" the transition and minimizing the energy at each step with respect to the other internal coordinates (8). Thus we may consider the problem of *application* of the model to ordinary sized organic molecules to be essentially solved. Further, experience has shown that the model itself is *valid* for significant classes of organic (and other) molecules. The question naturally arises, then, as to whether it might usefully be extended to bulk matter rather than just isolated molecules. It turns out that there are some very interesting problems in molecular solids that can be studied by the method.

Actually, it might be said that the first application of molecular mechanics calculations were to crystals. Crystal parameters and heats of sublimation of rare-gas solids and other simple molecules have long been used to help determine interatomic and intermolecular potentials (9). The problem of calculating crystal structures of molecular solids in general doesn't really

0-8412-0463-2/78/47-086-137$05.00/0

greatly extend the computational magnitude because of the restraints of the crystal symmetry. We do not address ourselves to this interesting and important application here. However, the structure and motion of defects in molecular solids does require the minimization of the conformational energy of mechanically stable and non-regular assemblages of molecules. The mechanically stable description (where a single minimum potential energy configuration of a large assembly is sought) is used to distinguish the problem from that of "Monte Carlo" or "molecular dynamics" calculations (where the object is to evaluate the potential energy or trajectories of a large number of random configurations for statistical averaging purposes).

An important example of the desirability of calculating the structure and mobility of defects in molecular solids lies in the area of polymer crystals. The onset of chain mobility without melting (which often can be treated as defect motion) with increasing temperature results in rather dramatic decreases in crystal stiffness (10). In the present work we describe a strategy which has been successful in handling the energy minimization of the large assembly of atoms characteristic of the environment of a defect in a molecular crystal. It was first applied to the properties of kinks in polyethylene (11) and has been subsequently applied to other problems (12,13) (see Figure 1) but the methodology has not been described in detail before.

A Strategy for Energy Minimization

The conformational energy of a molecular crystal may be written as the sum of the intra and intermolecular energies

$$U = U_{intra} + U_{inter} \quad . \tag{1}$$

The intramolecular energy is approximated as the sum of energies of deformation of valence bond lengths R_i, valence angles Θ_j and torsional angles ϕ_k and energies of non-bonded (Vander Waals) interactions at distances R_n,

$$U_{intra} = \Sigma \tfrac{1}{2} k_i^R (R_i - R_i^o)^2 + \Sigma \tfrac{1}{2} k_j^\Theta (\Theta_j - \Theta_j^o)^2 + \tfrac{1}{2} \Sigma V_k^\phi (1 + \cos 3\phi_k) + \Sigma V_{NB}^n (R_n) \quad . \tag{2}$$

The intermolecular energy is the sum of non-bonded energies associated with intermolecular atom distances, R_m

$$U_{inter} = \Sigma V_{NB}^m (R_m) \quad . \tag{3}$$

In the energy minimization of isolated organic molecules it has been found efficient to use a highly cooperative algorithm in which at each step of an iterative scheme all atoms move cooperatively towards the minimum. This is exemplified by multi-dimensional Newton-Raphson methods.

The Newton-Raphson Method (4). The energy in Eq. (2) may be expanded about a trial geometry through quadratic terms in the displacements from the trial structure ΔR_i, $\Delta\theta_j$, $\Delta\phi_k$ and ΔR_n. However, this function is highly redundant in these internal coordinates. The redundancy may be removed by expressing each internal coordinate displacement in terms of the three cartesian displacements $\Delta X_p^\alpha (\alpha = 1,2,3)$ of each of the appropriate atoms, p. The molecular energy becomes through quadratic terms in the ΔX's,

$$U = \Sigma A_p \Delta X_p^\alpha + \Sigma \tfrac{1}{2} C_{pq} \Delta X_p^\alpha \Delta X_q^{\alpha'} \quad (4)$$

where the constants A_p, C_{pq} now involve both the potential function constants of Eq. (2) and geometrical factors from the transformations from internal to cartesian coordinates. Eq. (4) can be differentiated with respect to the ΔX's to obtain a set of linear equations. From the solution, a set of displacements toward the minimum is obtained that is used to calculate a new trial structure. Iteration is continued until the A_p and the displacements approach zero. While there is no guarantee of convergence, it has been found that with the added feature of damping or attenuating large displacements the method converges reliably in a practical number of iterations ($\sim$2-10) and that the number of iterations tends to be independent of molecular size. However, the solution of the linear equations at each iteration associated with this method becomes impractical for large systems ($>\sim$60 atoms) because of the N^3 dependence of computation time.

A Hybrid Newton Raphson Relaxation Method. A section of one polymer chain containing or neighboring a conformational defect is about the right size to be handled well by the Newton-Raphson method (see Figure 2). This suggests preserving its basic utility and trying a "hybrid" technique. We find that a hybrid algorithm in which the internal degrees of freedom of each molecule are minimized by Newton-Raphson but in which the interactions between molecules are handled by a relaxation method is efficient. The method proceeds as follows:

(a) From an initial trial structure the energy (with respect to its internal coordinates and its position and orientation) of a selected molecule free to move in the field of its fixed neighbors is minimized by iterative Newton-Raphson.

(b) In turn each molecule in the assembly has its energy minimized in the field of the other fixed molecules.

(c) The total energy of the assembly is computed at the end of each complete cycle through the molecules.

(d) The cycling is repeated until convergence of the total energy is obtained.

In Step (a) above it is necessary to modify the Newton-Raphson algorithm used for free isolated molecules slightly to

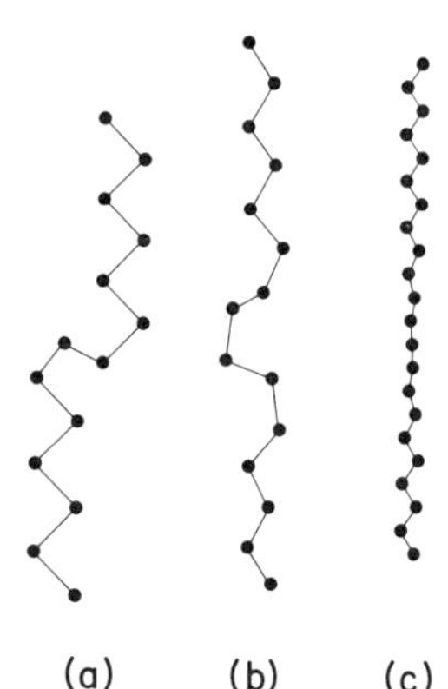

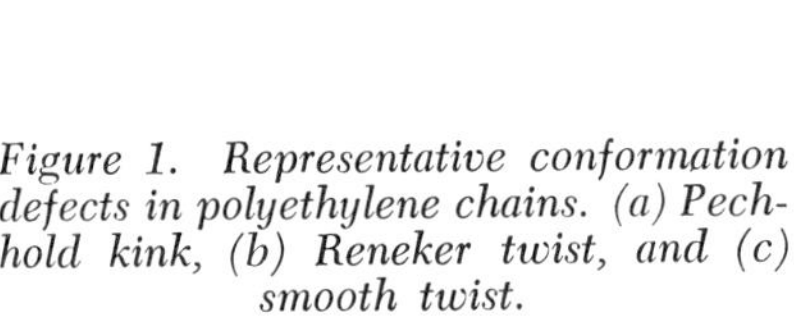
Figure 1. Representative conformation defects in polyethylene chains. (a) Pechhold kink, (b) Reneker twist, and (c) smooth twist.

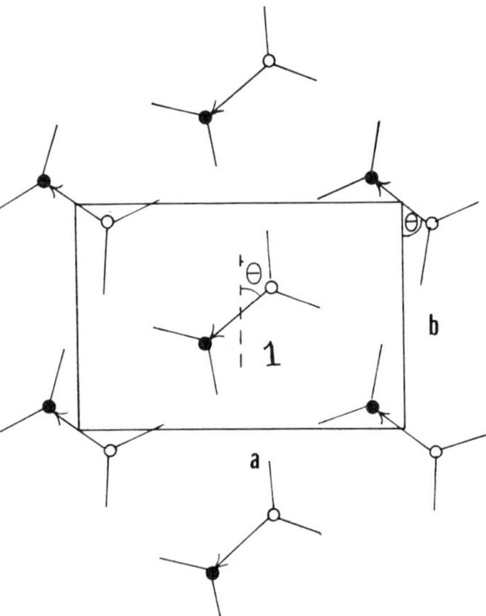

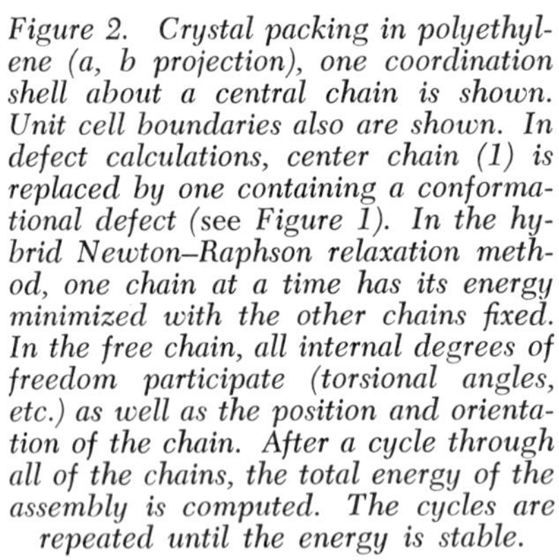
Figure 2. Crystal packing in polyethylene (a, b projection), one coordination shell about a central chain is shown. Unit cell boundaries also are shown. In defect calculations, center chain (1) is replaced by one containing a conformational defect (see Figure 1). In the hybrid Newton–Raphson relaxation method, one chain at a time has its energy minimized with the other chains fixed. In the free chain, all internal degrees of freedom participate (torsional angles, etc.) as well as the position and orientation of the chain. After a cycle through all of the chains, the total energy of the assembly is computed. The cycles are repeated until the energy is stable.

accommodate the intermolecular non-bonded interactions in crystals. In Eq. (2) above, R_n involves an intramolecular distance between two atoms, both of which are free to move toward the optimum position. In the relaxation method, R_m (Eq. (3)) represents an intermolecular interaction in which the atom in the chain selected for minimization moves but the other one is fixed. This situation requires an additional subroutine in the Newton-Raphson program. In the conventional Newton-Raphson program applied to single molecules a permanent table of the interactions contained in Eq. (2) is set up and used throughout the calculation. Because of the large number of intermolecular interactions this is abandoned in the intermolecular R_m subroutine. The list of interactions is generated dynamically in the R_m subroutine during each iteration from the lists of atoms coordinates.

Due to the large number of intermolecular interactions computation time is reduced by employing a distance cut-off beyond which interactions are ignored. We have used the non-bonded functions for carbon and hydrogen developed by Williams (14) and we employ in the R_m subroutine the same cut-off used by him in developing the functions.

The Banded-Matrix Approximation. The linear nature of the polymer chains allows an approximation which greatly speeds up the individual Newton-Raphson minimizations. As discussed above, the Newton-Raphson method results in a set of linear equations (represented by the coefficient or "$\underline{C}$" matrix in Eq. (4)). The operations in the Gauss reduction of the system increase as N^3 (where N is the number of variables). Atoms that are spatially removed from each will have small interaction coefficients, C_{pq}. In a large molecule many of the coefficients are small. In an extended linear molecule (such as encountered in the present context) numbering the atoms consecutively from one end to the other ensures that the large coefficients will be near diagnoal (see Figure 3). Those coefficients far off diagonal will represent spatially removed interactions and will be small. Thus we can approximate the $\underline{C}$ matrix as a banded one. The number of operations in reducing it increases as N for large N. This allows handling quite long chain segments in the crystal array with reasonable computation times.

Computational Results

The crux of the present method is how well the relaxation portion converges. Some experience has now been gained with it (11,12,13). Figure 4 shows results of calculations made in our laboratory (11,13) on the energies of three conformational defects in polyethylene crystals. These are a kink (15, 6), a Reneker twist (16) and a smooth twist (13) (see Figure 1). The first two have been proposed as stable point defects and the last one as a transition state for the motion that accomplishes

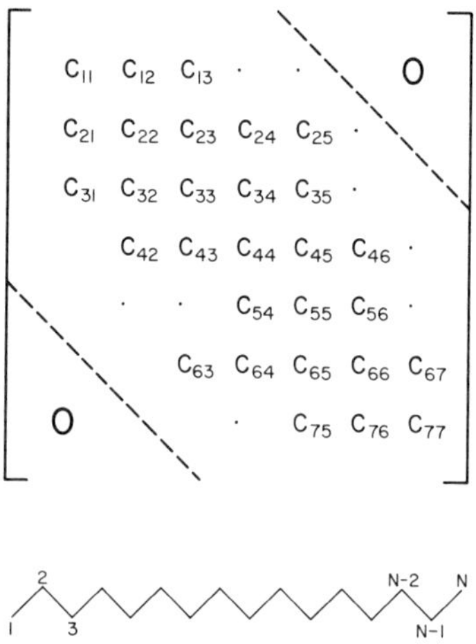

$$\begin{bmatrix} C_{11} & C_{12} & C_{13} & \cdot & \cdot & & 0 \\ C_{21} & C_{22} & C_{23} & C_{24} & C_{25} & \cdot & \\ C_{31} & C_{32} & C_{33} & C_{34} & C_{35} & \cdot & \\ & C_{42} & C_{43} & C_{44} & C_{45} & C_{46} & \cdot \\ & \cdot & \cdot & C_{54} & C_{55} & C_{56} & \cdot \\ & & C_{63} & C_{64} & C_{65} & C_{66} & C_{67} \\ 0 & & & \cdot & C_{75} & C_{76} & C_{77} \end{bmatrix}$$

1 2 3 N-2 N-1 N

Figure 3. Coefficient matrix in Newton–Raphson method. In a linear (nearly) extended chain, units far apart in numbering are far apart spatially and have small interaction coefficients. This orders the matrix in a banded manner that allows Gauss reduction to be greatly accelerated.

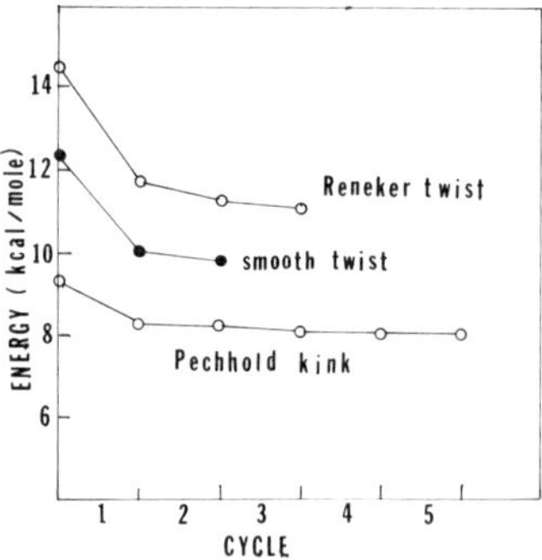

Figure 4. Convergence of the relaxation method. The energy of a crystalline array of polyethylene chains with the central one containing the defect indicated (relative to a nondefective array), see Figure 1. The Pechhold kink is in an array of two free coordination shells (6 +12 = 18 chains), each chain is $C_{14}H_{30}$. The other two defects are in an array of one free coordination shell plus a rigid second coordination shell, each chain is $C_{22}H_{68}$.

net rotation-translation of a crystal chain. Although the calculations have not been carried out to an extreme number of cycles it appears the convergence is satisfactory. The computation times are not excessive. For two flexible coordination shells about a polyethylene kink defect (19 chains, each $C_{14}H_{30}$) three complete relaxation cycles required about 30 min. computation time on a Univac 1108 system. Other assemblies should be roughly proportional to the number of chains and the number of units within each chain.

The consequences of the computed defect properties for polymer crystal behavior have been discussed in the literature cited and will not be gone into here. We suffice to conclude that it is possible to devise a computational strategy that allows the energy minimization of chain assemblies large enought to permit realistic and practical simulation of the energies and packings of defects in polymer crystals.

The U. S. Army Research Office (Durham) and the National Science Foundation provided financial support for the work reported here.

Literature Cited

1. Allinger, N. L., Adv. Phys. Org. Chem. (1976) 13, 1.
2. Wiberg, K. B., J. Am. Chem. Soc. (1965) 87, 1070.
3. Jacob, E. J., Thompson, H. and Bartell, L. S., J. Chem. Phys. (1967) 47, 3736.
4. Boyd, R. H., J. Chem. Phys. (1968) 49, 2574.
5. Lifson, S. and Warshel, A., J. Chem Phys. (1968) 49, 5116.
6. Scherr, H., Hägele, and Grossman, H. P., Colloid Polym. Sci. (1974) 252, 871.
7. Allinger, N. L., J. Amer. Chem. Soc. (1971) 93, 1637.
8. Wiberg, K. B. and Boyd, R. H., J. Amer. Chem Soc. (1972) 94, 8426.
9. Hirschfelder, J. O., Curtiss, C. F. and Bird, R. B., "Molecular Theory of Gases and Liquids" Wiley, New York (1954).
10. McCrum, N. G., Read, B. E. and Williams G., "Anelistic and Dielectric Effects in Polymeric Solids" Wiley, New York (1967).
11. Boyd, R. H., J. Polym. Sci.-Polym. Phys. Ed. (1975) 13, 2345.
12. Reneker, D. H., Fanconi, B. M. and Mazur, J., J. Appl. Phys. (1977) 48, 4032.
13. Mansfield, M. and Boyd, R. H., J. Polym Sci.-Polym. Phys. Ed. (1978) in press.
14. Williams, D. E., J. Chem Phys. (1967) 47, 4680.
15. Pechhold, W., Blasenbry, S. and Woerner, Colloid Polym. Sci. (1963) 189, 14.
16. Reneker, D. H., J. Polym. Sci. (1962) 59, 539.

RECEIVED September 7, 1978.

13

Multiple Time Step Methods and an Improved Potential Function for Molecular Dynamics Simulations of Molecular Liquids

W. B. STREETT[1] and D. J. TILDESLEY[2]

Science Research Laboratory, United States Military Academy, West Point, NY 10996

G. SAVILLE

Department of Chemical Engineering, Imperial College of Science, Prince Consort Road, London S.W.7., England

Among the important recent advances in computing methods for molecular dynamics simulations of condensed matter are the singularity free algorithm for rigid polyatomics, developed by Evans and Murad (1, 2), and the multiple time step (MTS) method, developed by Streett, et al. (3). In the method of Evans and Murad, the equations of rotational motion are expressed in terms of four parameters, χ, η, ξ, ζ, called quaternions, defined by

$$\chi=\mathrm{Cos}(\theta/2)\,\mathrm{Cos}((\psi+\phi)/2)$$

$$\eta=\mathrm{Sin}(\theta/2)\,\mathrm{Cos}((\psi-\phi)/2)$$

$$\xi=\mathrm{Sin}(\theta/2)\,\mathrm{Sin}((\psi-\phi)/2)$$

$$\zeta=\mathrm{Cos}(\theta/2)\,\mathrm{Sin}((\psi+\phi)/2),$$

where θ, ψ and ϕ are the Euler angles defined by Goldstein (4). They report significant improvements in accuracy and efficiency over methods based on the equations of Euler, Lagrange or Hamilton, using Euler angles. The MTS method of Streett, et al., based on the use of two or more time steps of different lengths to integrate the equations of motion in systems governed by continuous potential functions, has resulted in increases in computing speed by factors of three to eight over conventional methods (3). The original MTS paper describes in detail how the method is applied to systems of spherical molecules. In this paper we give similar details of the application of the MTS method to systems of rigid polyatomic molecules of arbitrary shape, using the algorithm of Evans and Murad. We also

[1] Current address: Chemical Engineering Department, Cornell University, Ithaca, NY 14853.
[2] Current address: Physical Chemistry Laboratory, South Parks Road, Oxford OX1 3QZ, England.

0-8412-0463-2/78/47-086-144$05.00/0

describe a modified site-site potential, the shifted force potential, that gives better accuracy and stability than the usual truncated potential.

Site-Site Potentials For Molecular Liquids

Since site-site potentials (also called atom-atom potentials) have been widely used in computer simulation and theoretical studies of molecular liquids, we describe the application of MTS methods to a model potential of this type. MTS methods are completely general, and can be applied equally well to other types of potential functions. In site-site models the potential energy between a pair of molecules is the sum of the potential energies between pairs of sites on different molecules. These sites are usually atoms or groups of atoms, which interact with sites on other molecules through spherically symmetric potential functions. Thus the potential energy between molecules 1 and 2 takes the form

$$U(r_{12}\omega_1\omega_2)=\sum_{\alpha\beta} u_{\alpha\beta}(r_{\alpha\beta}),$$

where ω_i is the set of angles defining the orientation of molecule i, $u_{\alpha\beta}(r_{\alpha\beta})$ is the potential energy between site α on molecule 1 and site β on molecule 2 (a function only of the site-site distance $r_{\alpha\beta}$), and the summation is over all sites in each molecule. The methods described herein are applicable to models governed by site-site potentials that are continuous functions of r.

Truncated Potentials. In the interest of computing efficiency, site-site potentials are truncated at a cutoff distance, r_c, usually taken to be about 2.5 times the effective "diameter" of the site. Interactions beyond this distance are neglected, and corrections to the calculated thermodynamic properties, due to the neglected "tail", are added using a simple perturbation scheme (5).

Errors inherent in the use of truncated potentials in molecular dynamics simulations have usually been overlooked, or at least neglected. These errors arise because of the impulse felt by two molecules whose separation r crosses the boundary $r=r_c$ in successive time steps. There is a gain or loss of potential energy that is not balanced by an equal and opposite change in kinetic energy; total energy is not conserved, and the calculated trajectories depart slightly from true Newtonian trajectories. (Numerical round off causes similar errors.) In simulations of homogeneous fluids there are, on average, as many crossings of the boundary in one direction as the other, so there is no appreciable drift in the total energy and no significant error in the calculated equilibrium properties; however, in simulations of non-homogeneous systems (e.g., gas-liquid interface) these impulses can be a significant source of error and instability. In addition, the accumulation of small errors in particle trajectories is likely to result in significant errors in calculations of time-dependent phenomena that are based on correlations of properties at long times, such as the "long time tails" of the velocity autocorrelation function (6). For reasons to be made clear in the next section, truncated potentials have an adverse effect on energy conservation in some MTS methods. These problems can be overcome by simply increasing the cutoff distance r_c to a point where the interactions omitted are truly negligible; however, since the number of pair interactions calculated in the simulation increases approximately as r_c^3, this remedy is expensive. As an alternative, we have developed a modified potential, called the shifted force potential, that provides acceptable accuracy and energy conservation without increases in r_c.

The Shifted Force Potential. When two sites interact through a spherically symmetric potential function, they exert upon each other equal and opposite forces that act along the vector $\underset{\sim}{r}$ between them. (A wavy underline indicates a vector quantity.) The magnitude of the force is given by $F=-(du/dr)$. In Figure 1(a) the potential energy and force between two sites are plotted against distance for a typical truncated potential. It is the discontinuity in the force at $r=r_c$ that is the source of error in molecular dynamics simulations. To reduce the error we use a modified force, F_m, formed by subtracting from the usual expression for the force a constant, H, equal to the magnitude of this discontinuity. The modified force is

$$\begin{aligned} F_m &= -(du/dr)-H, \quad r \le r_c, \\ F_m &= 0, \quad r > r_c, \end{aligned} \tag{1}$$

where

$$H=-(du/dr)_{r=r_c}.$$

The modified potential is defined in terms of this force:

$$u_m(r) \equiv -\int_{\infty}^{r} F_m dr. \tag{2}$$

This gives

$$\begin{aligned} u_m(r) &= u(r)-Hr+G, \quad r \le r_c, \\ u_m(r) &= 0, \quad r > r_c, \end{aligned} \tag{2a}$$

where G is a constant given by

$$G=Hr_c-u(r_c).$$

The modified force and potential energy are plotted in Figure 1(b). Both potential energy and force now go smoothly to zero at $r=r_c$. The modification to the force is a constant equal to the magnitude of the original discontinuity at r_c, hence the name shifted force potential. The modification to the potential energy is a straight line $u_o(r)=-Hr+G$, with slope $-H$ and with u and r intercepts G and r_c, respectively. These modifications are applied only for $r \leq r_c$.

In simulations of heteronuclear polyatomics several different site-site potentials are used. The cutoff distance, r_c, used to modify each potential according to equations (1) and (2a), should be at least as large as the distance inside which that potential is always calculated during the simulation. Depending on the conventions used, these distances can be different for different site-site potentials. (See Figure 3.)

The small changes to the force and potential energy, resulting from the change to a shifted force potential function, provide improved accuracy and stability in computer simulations, with negligible changes to structure and time correlation functions, at short to moderate times, calculated from the usual truncated potential. Corrections to calculated thermodynamic properties to account for modifications to the potential can be calculated by a perturbation method similar to that used for the long tail corrections. This matter will be discussed in detail in a separate paper.

Multiple Time Step Methods

Spherical Molecules. Before describing the application of MTS methods to molecular liquids, we outline brifely how they are applied to spherical molecules; complete details can be found in the original paper (3). The essence of the method is the separation of the total instantaneous force acting on each molecule into a primary force, $\mathbf{F}_p$, produced by a cage of nearest neighbors, and a secondary force, $\mathbf{F}_s$, produced by the remaining neighbors within the cutoff distance, and the use of time steps of different lengths to calculate the time evolution of these forces. In general, $F_p > F_s$, and similar inequalities hold for time derivatives of these forces. This means that the motion of a molecule is dominated by a strong, rapidly varying primary force, produced by a cage of nearest neighbors, while the smaller secondary force changes more slowly with time. In conventional molecular dynamics methods, no distinction is made between primary and secondary forces; the forces between all pairs within the cutoff distance are recalculated every time step, and it is this summation over interacting pairs that consumes most of the computing time. In the MTS method, only the primary forces are recalculated at every step, while time steps five to twenty times longer are used to predict the time evolution of secondary forces. In general, the number of pri-

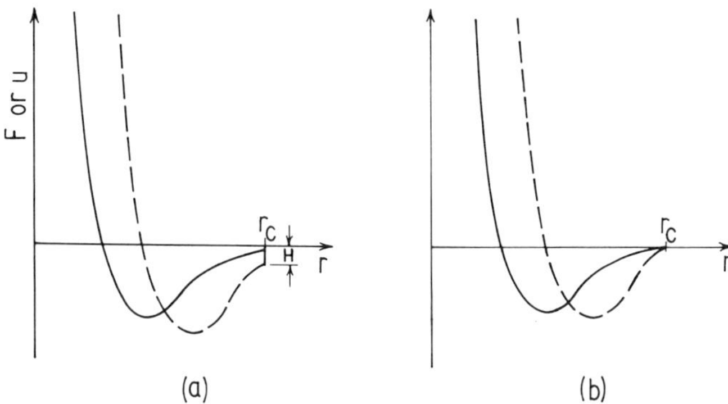

Figure 1. Potential energy u *(solid lines) and force* F *(dashed lines) as a function of site–site distance* r *for truncated (a) and shifted-force (b) potential functions. The modified force in (b) is formed by shifting the force–distance curve in (a) upward a distance* H, *equal to the magnitude of the discontinuity at* $r = r_c$. *The modified potential energy in (b) is then defined by Equations 1 and 2.*

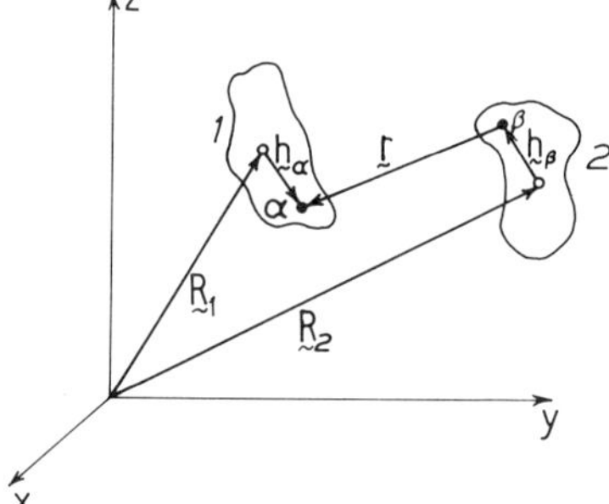

Figure 2. The vector r *from site* β *of molecule 2 to site* α *of molecule 1, in the space-fixed axis system, is given by Equation 6. Open circles are molecular centers. To apply the Taylor series MTS methods (Equation 3), one or more time derivatives of* r *must be calculated.*

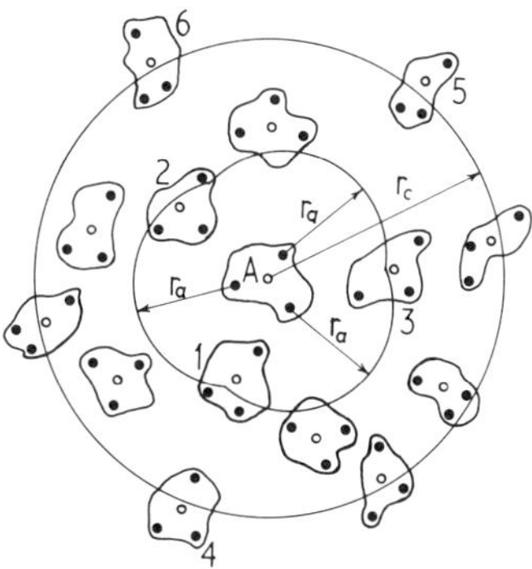

Figure 3. Diagram illustrating the conventions chosen for excluding interactions beyond the cutoff distance and for distinguishing between primary and secondary forces. Open circles are molecular centers and solid circles are additional sites. All sites on molecules 1, 2, and 3 contribute to the primary forces on the central molecule A, because each has at least one site within a distance r_a *from a site of A. Molecules 4, 5, and 6 do not interact with A because their centers lie beyond the cutoff distance,* r_c, *from the center of A. All site–site interactions between molecule A and the remaining (unlabeled) molecules contribute to the secondary forces on A.*

mary neighbors is about an order of magnitude less than the number within the cutoff distance, so the average number of pair interactions explicitly calculated at each time step is greatly reduced.

At time t_0, the primary and secondary forces, $\underset{\sim}{F}_p(t_0)$ and $\underset{\sim}{F}_s(t_0)$, are calculated for each molecule. Concurrently, the time derivatives, $\underset{\sim}{F}_s'(t_0)$, $\underset{\sim}{F}_s''(t_0)$,..., of the secondary force are calculated, and lists of the primary neighbors are compiled for each molecule. At each of the next n-1 time steps (where, typically, n=10) the primary forces are recalculated in the usual way, using the list of primary neighbors compiled at t_0, while the secondary forces are calculated from a Taylor series, using time derivatives calculated at t_0:

$$\underset{\sim}{F}_s(t_0+k\Delta t)=\underset{\sim}{F}_s(t_0)+\underset{\sim}{F}_s'(t_0)\frac{k\Delta t}{1!}+\cdots+\underset{\sim}{F}_s^{(m)}(t_0)\frac{(k\Delta t)^m}{m!}+\cdots; \quad k=1,2,\cdots,n-1. \tag{3}$$

Here Δt is the time step, and we shall refer to the Taylor series truncated after term m as the mth order Taylor series. An alternative first order prediction scheme, based on the finite difference formula

$$\underset{\sim}{F}_s'=\frac{\underset{\sim}{F}_s(t_0+\Delta t)-\underset{\sim}{F}_s(t_0)}{\Delta t} \tag{4}$$

has also been successfully tested. In this method, which we call the linear extrapolation MTS method, the primary and secondary forces between all pairs within the cutoff distance are calculated at each of the first two steps in each block of n steps, and a linear extrapolation of the secondary forces is carried out over the next n-2 steps. When the linear extrapolation method is applied to molecular liquids, torques must be separated into primary and secondary components and treated in the same way as forces. This method is similar to the first order Taylor series method (equation 3 with two terms), but it is simpler to use and requires less storage. When a truncated potential is used, the linear extrapolation method gives better energy conservation because it includes the effects of molecular pairs crossing the boundary $r=r_c$ between the first two time steps in each block, and uses this information in the extrapolation of secondary forces over the next n-2 steps. Taylor series methods, on the other hand, include the inherent assumption that no crossings occur in each block of n steps. The use of shifted force potentials with these methods results in improved accuracy and stability. Comparisons of computing speeds and storage requirements for various MTS methods are given in the next section.

Expressions for the time derivatives $\underset{\sim}{F}_s'$, $\underset{\sim}{F}_s''$,..., in equation (3) have been published (3). The first of these is

$$\underset{\sim}{F}'=A\underset{\sim}{r}'+B(\underset{\sim}{r}\cdot\underset{\sim}{r}')\underset{\sim}{r}, \tag{5}$$

where A and B are functions only of the scalar r. In equation (5) the time dependence of the force between a pair of molecules is thus contained in the time derivative $\underset{\sim}{r}'$ of the vector separating them. Higher derivatives of $\underset{\sim}{F}$ contain higher derivatives of $\underset{\sim}{r}$. The vector $\underset{\sim}{r}_{12}$ between two spherical molecules is a simple vector sum: $\underset{\sim}{r}_{12}=\underset{\sim}{r}_1-\underset{\sim}{r}_2$, where $\underset{\sim}{r}_i$ is the position vector of molecule i in a space-fixed frame. This expression can be differentiated to give $\underset{\sim}{r}_{12}'=\underset{\sim}{r}_1'-\underset{\sim}{r}_2'$, $\underset{\sim}{r}_{12}''=\underset{\sim}{r}_1''-\underset{\sim}{r}_2''$, etc. The time derivatives of $\underset{\sim}{r}_{12}$ are simple vector sums of the velocities, accelerations, and higher time derivatives of the center of mass positions.

Molecular Liquids. In the case of rigid polyatomic molecules governed by site-site potentials, the situation is slightly more complicated, because forces act on sites other than the center of mass, and rotational motion must be taken into account. In most schemes for solving the equations of motion of polyatomics, the motion is separated into translation of the center of mass and rotation about an axis through the center of mass. In the summation of forces step in the simulation the total force acting on each site of each molecule is calculated by a summation over all molecular pairs within the cutoff distance. Once these forces are known, the torque on each molecule is calculated from a sum of vector cross products of the form

$$\underset{\sim}{T}=\sum_{\alpha}(\underset{\sim}{h}_{\alpha}\times\underset{\sim}{F}_{\alpha}), \tag{5a}$$

where $\underset{\sim}{T}$ is the torque, $\underset{\sim}{h}_{\alpha}$ is the vector from the center of mass to site α, and $\underset{\sim}{F}_{\alpha}$ the force on site α. For site-site models, the MTS method is applied only to the calculation of site-site forces; the torques are then calculated in the usual way.

Figure 2 shows the relation between the site-site vector $\underset{\sim}{r}$ connecting sites α and β, the position vectors $\underset{\sim}{R}_1$ and $\underset{\sim}{R}_2$ of molecules 1 and 2, and the body-fixed site vectors $\underset{\sim}{h}_{\alpha}$ and $\underset{\sim}{h}_{\beta}$. The vector $\underset{\sim}{r}$ can be expressed as

$$\underset{\sim}{r}=\underset{\sim}{R}_1-\underset{\sim}{R}_2+\underset{\sim}{h}_{\alpha}-\underset{\sim}{h}_{\beta}. \tag{6}$$

Differentiating gives

$$\begin{aligned}\underset{\sim}{r}'&=\underset{\sim}{R}_1'-\underset{\sim}{R}_2'+\underset{\sim}{h}_{\alpha}'-\underset{\sim}{h}_{\beta}',\\ \underset{\sim}{r}''&=\underset{\sim}{R}_1''-\underset{\sim}{R}_2''+\underset{\sim}{h}_{\alpha}''-\underset{\sim}{h}_{\beta}'', \text{ etc.}\end{aligned} \tag{7}$$

To apply the MTS method to molecular liquids we need to calculate the derivatives on the right side of equation (7). The derivatives $\underset{\sim}{R}^{(m)}$ of the center of mass position vectors are routinely calculated to order m=4 or 5 in molecular dynamics programs based on predictor-corrector algorithms of the type due to Gear (7). To calculate the derivatives $\underset{\sim}{h}'_\alpha$, $\underset{\sim}{h}''_\alpha$, etc., we take advantage of the elegant features of the method of quaternions. In this method, the equations of rotational motion are solved in a body-fixed system of axes (the principal axis system). Any vector $\underset{\sim}{V}_P$ in the principal axis system is transformed into a space-fixed vector $\underset{\sim}{V}_A$ by the transformation

$$\underset{\sim}{V}_A=\underline{\underline{E}}\cdot\underset{\sim}{V}_P,$$

where $\underline{\underline{E}}$ is the rotation matrix

$$\underline{\underline{E}}=\begin{vmatrix} -\xi^2+\eta^2-\zeta^2+\chi^2 & -2(\xi\eta+\zeta\chi) & 2(\eta\zeta-\xi\chi) \\ 2(\zeta\chi-\xi\eta) & \xi^2-\eta^2-\zeta^2+\chi^2 & -2(\xi\zeta+\eta\chi) \\ 2(\eta\zeta+\xi\chi) & 2(\eta\chi-\xi\zeta) & -\xi^2-\eta^2+\zeta^2+\chi^2 \end{vmatrix} \quad (8)$$

Hence the vector $\underset{\sim}{h}_\alpha$ is given by

$$\underset{\sim}{h}_\alpha=\underline{\underline{E}}\cdot\underset{\sim}{b}_\alpha,$$

where $\underset{\sim}{b}_\alpha$ is the vector from the center of mass of the molecule to site α, in the principal axis system. Now $\underset{\sim}{b}_\alpha$ is time invariant, so the derivatives of h_α are given by

$$\begin{aligned} \underset{\sim}{h}'_\alpha&=\underline{\underline{E}}'\cdot\underset{\sim}{b}_\alpha, \\ \underset{\sim}{h}''_\alpha&=\underline{\underline{E}}''\cdot\underset{\sim}{b}_\alpha, \text{ etc.} \end{aligned} \quad (9)$$

The derivatives $\underline{\underline{E}}'$, $\underline{\underline{E}}''$, etc. are obtained from term by term differentiation of equation (8). The first derivatives of the quaternions can be calculated from the principal angular velocity vector, $\underset{\sim}{\omega}$, in the predictor step, using the following expression,

$$\begin{vmatrix} \xi' \\ \eta' \\ \zeta' \\ \chi' \end{vmatrix} = \tfrac{1}{2} \begin{vmatrix} -\zeta & -\chi & \eta & \xi \\ \chi & -\zeta & -\xi & \eta \\ \xi & \eta & \chi & \zeta \\ -\eta & \xi & -\zeta & \chi \end{vmatrix} \begin{vmatrix} \omega_x \\ \omega_y \\ \omega_z \\ 0 \end{vmatrix} \quad (10)$$

Higher derivatives can be obtained by successive differentiation of equation (10), using the derivatives $\underset{\sim}{\omega}'$, $\underset{\sim}{\omega}''$, etc., which are routinely calculated in predictor-corrector integration schemes. Thus equations (8)-(10) provide the means to calculate the derivatives $\underset{\sim}{h}'_\alpha$, $\underset{\sim}{h}''_\alpha$, etc., in equation (7). Once the derivatives $\underset{\sim}{r}'$, $\underset{\sim}{r}''$, etc., for the site-site distances are known, the MTS method is applied to calculation of the forces in much the same manner as in the case of spherical molecules.

Certain thermodynamic properties, such as pressure and internal energy, must be separated into contributions due to primary and secondary interactions, and calculated in a manner analogous to that used for the forces (3).

See appendix for outline of the molecular dynamics program.

Computing Time and Storage Requirements

General. The results obtained for the tetrahedral model used here are only a single example of the application of MTS methods to molecular liquids. Results obtained in other cases will depend on many factors, including molecular shape, type of intermolecular potential function, and numerical method used to solve the equations of motion.

We describe here increases in computing speed and storage requirements associated with adapting a Fortran program for 108 methane (CH_4) molecules from conventional molecular dynamics to several variations of the MTS method. The program uses the singularity free algorithm of Evans and Murad (2), and a site-site potential of the form

$$u(r)=Ar^{-6}+B\exp(Cr),$$

where A, B, C are constants which depend on the type of site-site interaction (carbon-carbon, carbon-hydrogen, or hydrogen-hydrogen). Each molecule has five sites (atoms), so there are 25 site-site interactions per pair of molecules. Distances are expressed in units of σ, the effective "diameter" of a methane molecule (4.01Å in this case). For the cutoff r_c we use the distance between molecular centers: all interactions between sites of a pair of molecules are calculated if $R_{12}\leq r_c$, none if $R_{12}>r_c$. To distinguish between primary and secondary forces we use the minimum site-site distance, r_{min}, between a pair of molecules: if $r_{min}\leq r_a$ (where r_a is typically about 1.2σ) all site-site interactions for that pair contribute to primary forces; if $r_{min}>r_a$, and $R_{12}\leq r_c$, all site-site interactions contribute to secondary forces. These conventions are illustrated in Figure 3.

Computer Storage. Increases in computer storage to accommodate MTS methods are given in units of the number of molecules, N, and the number of sites, N_s, in the system (e.g., in a system of N=108 methane molecules, N_s=540). In every case the storage required for the list of primary neighbors is $\sim$6N. For MTS methods based on the Taylor series expansion, equation (3), there is an added requirement of $\sim 5N_s$ words for each term in equation (3). Thus for a 2nd order Taylor series (three terms) the storage increases by $\sim 15N_s$. The linear extrapolation method (equation 4) requires only about 25N additional words of storage. Examples of storage requirements for several MTS programs are given in the table below.

Computing Time. In molecular dynamics simulations most of

the computing time is spent in calculating interactions between pairs of molecules that lie within a specified cutoff distance, hence computing speed varies inversely with the average number of pair interactions calculated per time step. Let us consider as a base for comparison a conventional molecular dynamics program in which this average number is N_t. Using MTS methods, the simulation is carried forward in blocks of n steps where, during the first one or two steps the usual N_t interactions are calculated, together with the derivatives in equation (3) or (4), while during the remaining steps the interactions between a lesser number of primary neighbors, N_p, are calculated. N_t is determined by the cutoff distance r_c, and N_p by the distance r_a used to distinguish between primary and secondary neighbors. For the tetrahedral model described here we have used $\sigma \leq r_a \leq 1.3\sigma$ and $r_c = 2.5\sigma$, which give $0.2 \leq N_p/N_t \leq 0.4$.

The computing speed of a particular MTS program is a complicated function of the parameters r_a (the distance separating primary and secondary neighbors), n (the number of time steps in each block) and m (the order of the Taylor series used, equation 3). The speed increases with increasing n and with decreasing r_a and m. An acceptable combination of these parameters is one that results in adequate energy conservation, which, for the purposes of this example, we have arbitrarily chosen to be 0.05 per cent per 1000 time steps. Listed below are examples of the computing speed and storage requirements for a conventional molecular dynamics program and several MTS programs for a system of 108 tetrahedral molecules, all based on the singularity free algorithm of Evans and Murad. We used a time step of 2.4×10^{-15} sec, the maximum time step which gives satisfactory energy conservation in the conventional program. Computing speeds are in units of the speed of the conventional program. (This speed is about 1.8 time steps per minute of computing time on the Honeywell H-6080 computer at the United States Military Academy. We estimate that it is equivalent to about 20 to 25 time steps per minute on a CDC 7600 or IBM 370.)

Program	r_a	n	Storage (10^3 words)	Computing Speed
Conventional, without MTS	--	--	30	1
MTS, linear extrapolation	1.3	6	33	2.4
MTS, 2nd order Taylor series	1.1	6	38	2.6
MTS, 2nd order Taylor series	1.2	10	38	3.2
MTS, 2nd order Taylor series	1.2	12	38	3.2
MTS, 3rd order Taylor series	1.0	6	41	2.5
MTS, 3rd order Taylor series	1.2	15	41	3.4

These figures show that the MTS method increases computational speeds by factors of three or more when applied to a tetrahedral atom-atom model. Evans and Murad (2) report comparable increases for the singularity free algorithm over previous methods based on other algorithms; hence the combination of MTS methods with their algorithm can be expected to increase computing speeds by about an order of magnitude.

Our experience with simulations of diatomic molecules suggests that the advantage in computing speed of MTS methods increases with the asymmetry of the molecular model. It is therefore likely that greater relative increases in speed can be realized in simulations of systems of more highly asymmetric molecules than those described here. In addition, the relative increase in speed can be expected to increase as the number of molecules in the sample increases.

We have demonstrated the power of MTS methods in molecular dynamics simulations of both monatomic and polyatomic models. In another report in this volume, J. M. Haile describes how MTS methods have been used to develop an efficient algorithm for calculating three-body forces in systems of spherical molecules.

We are indebted to D. J. Evans and S. Murad for stimulating discussions of the singularity free algorithm and the use of quaternions in molecular dynamics simulations. We acknowledge a generous grant of computing time from the Academic Computer Center of the United States Military Academy, West Point, New York, and financial support from NSF grant ENG 76-82101.

Appendix

The following outline describes how MTS methods are combined with predictor-corrector integration schemes in a molecular dynamics program based on the singularity free algorithm of Evans and Murad (2). We consider a system of N polyatomic molecules governed by a site-site potential function, integrated by a predictor-corrector method of order ℓ, using a Taylor series MTS method of order m in which the simulation is carried forward in blocks of n time steps. Typical values of these parameters are: N=108, 256, 500,···, ℓ=4 or 5, m=2 or 3, n=10 to 15. The major sections of the program are as follows.

1. Initial Section. Assign initial positions and velocities to all molecules. The position of a molecule is determined by its center of mass position vector $\underset{\sim}{R}$ and four quaternions (χ, η, ξ, ζ). We have found it convenient to use a face centered cubic lattice, with random molecular orientation, as the initial configuration. (Note that the quaternions are subject to the constraint $\chi^2+\eta^2+\xi^2+\zeta^2=1$.) Assign initial translational velocities $\underset{\sim}{R}'$ and body-fixed rotational velocities $\underset{\sim}{\omega}$; the directions of these vectors can be chosen randomly, but the distribution of magnitudes should correspond to the desired temperature of the simulation. Calculate first derivatives of the quaternions from equation (10). Calculate the derivatives $\underset{\sim}{R}''$ and $\underset{\sim}{\omega}'$ from the forces and torques for the initial configuration, using the differential equa-

tions of motion (equations A2), as in section 3. Set higher derivatives of $\underset{\sim}{R}$, $\underset{\sim}{\omega}$ and the quaternions (to order ℓ) equal to zero.

2. Predictor Section. Calculate the first derivatives of the quaternions from equation (10), and higher derivatives (to order ℓ) from repeated differentiation of that equation. Using a Taylor series of order ℓ, predict values of $\underset{\sim}{R}$, $\underset{\sim}{\omega}$ and the quaternions, and the first ℓ derivatives of these quantities, at time $t+\Delta t$ according to

$$\begin{aligned} R(t+\Delta t) &\simeq R(t)+R'(t)\frac{\Delta t}{1!}+\cdots+R^{(\ell)}(t)\frac{\Delta t^{\ell}}{\ell!} \\ R'(t+\Delta t) &\simeq R'(t)+R''(t)\frac{\Delta t}{1!}+\cdots+R^{(\ell)}(t)\frac{\Delta t^{\ell-1}}{(\ell-1)!} \\ &\vdots \\ R^{(\ell)}(t+\Delta t) &\simeq R^{(\ell)}(t), \text{ etc.} \end{aligned} \tag{A1}$$

(This Taylor series should not be confused with equation (3), which is used in section 3.) Apply the constraint $\chi^2+\eta^2+\xi^2+\zeta^2=1$, to the predicted quaternions, rescaling each one by the factor $1/(\chi^2+\eta^2+\xi^2+\zeta^2)^{\frac{1}{2}}$.

3. Evaluation Section. Using predicted positions and orientations, calculate the total force on each site of each molecule. Depending on whether this time step is the first time step (a) or a subsequent time step (b) in a block of n steps, proceed as follows with the MTS method.
   (a) First Time Step. Calculate forces by a summation over all pairs within the cutoff distance r_c, and separate the force on each site into primary ($\underset{\sim}{F}_p$) and secondary ($\underset{\sim}{F}_s$) components (see Figure 3). Calculate the derivatives $\underset{\sim}{F}_s'$, $\ldots$, $\underset{\sim}{F}_s^{(m)}$ as outlined in the main text. For each molecule i compile a list of the remaining molecules that contribute to the primary forces on i. (Note that this list need only include molecules j for $j>i$, since the i-j and j-i interactions are identical.) Separate contributions to the thermodynamic properties (pressure, energy, etc.) into primary and secondary components, and calculate the time derivatives of the latter, using the expressions given in reference 3.
   (b) Subsequent Time Step. Calculate the primary force on each site and the primary-neighbor contributions to the thermodynamic properties by a summation over all pairs included in the list of primary neighbors compiled in the first time step. Calculate secondary forces from equation (3) and secondary-neighbor contributions to the thermodynamic properties from similar equations (see reference 3).

   Add the primary and secondary components of the contributions to the thermodynamic properties. Add the primary and secondary forces for cach site, and calculate the exact force and torque, in the space-fixed frame, on each molecule. Transform space-fixed torques, $\underset{\sim}{T}$, into torques in the principal-axis system, $\underset{\sim}{T}_P$, using $\underset{\sim}{T}_P=\underline{\underline{E}}^{-1}\cdot\underset{\sim}{T}$, where $\underline{\underline{E}}^{-1}$ is the inverse of the matrix in equation (8). (The matrix is orthogonal.) Calculate the exact translational and rotational accelerations, $\underset{\sim}{R}''$ and $\underset{\sim}{\omega}'$, from the differential equations of motion:

$$\underset{\sim}{R}''=\underset{\sim}{F}/M$$

$$\omega'_\gamma=T_{P_\gamma}/I_{P_\gamma}\ , \qquad \gamma=x,y,z, \tag{A2}$$

where M is the molecular mass and I_{P_x}, I_{P_y}, I_{P_z} the principal moments of inertia (4). Calculate exact values of the first derivatives of the quaternions from equation (10).

4. Corrector Section. Correct the predicted values of $\underset{\sim}{R}$ and $\underset{\sim}{\omega}$, and their derivatives, and the predicted values of the quaternions (but not their derivatives), according to

$$\underset{\sim}{R}^{(j)}_{corr}=\underset{\sim}{R}^{(j)}_{pred}+\frac{A_j\cdot\Delta\underset{\sim}{R}''}{(\Delta t)^j/j!}, \qquad j=0,1,\cdots,\ell,$$

$$\underset{\sim}{\omega}^{(j)}_{corr}=\underset{\sim}{\omega}^{(j)}_{pred}+\frac{A_j\cdot\Delta\underset{\sim}{\omega}'}{(\Delta t)^j/j!}, \qquad j=0,1,\cdots,\ell, \tag{A3}$$

$$\xi_{corr}=\xi_{pred}+\frac{A_1\cdot\Delta\xi'}{\Delta t/1!}, \qquad \text{etc.},$$

where $\Delta\underset{\sim}{R}''$, $\Delta\underset{\sim}{\omega}'$ and $\Delta\xi'$ are the differences between the predicted values from section 2 and the exact values from section 3. The constants A_j for the predictor-corrector method can be found in table 9.1 of reference 7.

5. Return to Predictor Section.

To use the linear extrapolation MTS method, the same procedure is followed, except that in the second time step of each block of n steps, the primary and secondary forces are again calculated separately, using the list of primary neighbors compiled in the first time step, and values of $\underset{\sim}{F}'_s$ from equation (4) are used in a first order Taylor series (equation 3) during the next n-2 time steps. If the torques are separated into primary and secondary components, and treated in the same manner as the forces, equations (3) and (4) can be applied to the total force on each molecule, rather than to the force on each site, resulting in a substantial savings in storage. (A similar savings in storage can be realized in the Taylor series MTS methods if the torques are separated into primary and secondary components, $\mathbf{T}_p$ and $\mathbf{T}_s$, and derivatives of $\mathbf{T}_s$ calculated from derivatives of equation (5a); however, the savings in storage will be offset by the increase in computing time required to evaluate these derivatives. In this case it is more efficient to sum the forces on the sites in a double loop over all relevant molecular pairs, and then to calculate the torques from equation (5a) in a single loop over all molecules.)

Literature Cited

1. Evans, D. J., Molec. Phys. (1977) 34, 317.
2. Evans, D. J. & Murad, S., Molec. Phys. (1977) 34, 327.
3. Streett, W. B., Tildesley, D. J. and Saville, G., Molec. Phys. (1978) in press.
4. Goldstein, H., "Classical Mechanics", Chaps. 4 & 5, Addison-Wesley, Reading, Massachusetts, 1971.
5. McDonald, I. R. & Singer, K., Disc. Faraday Soc. (1967) 43, 40.
6. Alder, B. J. & Wainwright, T. E., Phys. Rev. (1970) A1, 18.
7. Gear, C. W., "Numerical Initial Value Problems in Ordinary Differential Equations," pp. 148-54, Prentice-Hall, Englewood Cliffs, New Jersey, 1971.

RECEIVED September 7, 1978.

14

Optimization of Sampling Algorithms in Monte Carlo Calculations on Fluids

JOHN C. OWICKI

Department of Chemistry, Stanford University, Stanford, CA 94305

The analysis of simple model classical fluids by Monte Carlo (MC) techniques now is fairly routine. This is not so in more complicated systems, where computational expense may limit the precision of the results or the number of calculations performed. Examples include liquid water, dilute solutions, and systems with phase boundaries. Even in "simple" systems, some properties (notably free energies) are very difficult to calculate precisely.

The scope of feasible calculations can be expanded by reducing the cost of computing. Since computer technology continues to improve, this factor is significant. MC standard errors are asymptotically proportional to the reciprocal of the square root of the length of the computation, however, so improvements in precision will not be as great as drops in computational costs.

Another approach is to improve sampling procedures, since it is a maxim for MC calculations in general that a properly designed sampling algorithm can reduce statistical errors dramatically (1). Therefore, this chapter deals with the optimization of sampling algorithms for MC calculations on bulk molecular systems, particularly fluids. Part I is a discussion of some developments in the mathematical theory of MC sampling using Markov chains. The scope is purposely limited, and important topics such as umbrella sampling and sampling in quantum mechanical systems are given short shrift, because they are well-reviewed elsewhere (2,3). Nor do we discuss the work of Bennett (4), which deals primarily with efficient data analysis in MC free energy calculations, not with the sampling algorithms *per se*. Part II of the chapter is a generalization of a sampling algorithm (5) which we have developed for MC calculations on dilute solutions, with some calculations to illustrate its usefulness.

I. Theoretical Developments

The Metropolis Algorithm. For reference purposes and to establish notation, the standard sampling algorithm of Metropolis

0-8412-0463-2/78/47-086-159$05.00/0

et al. (6) is summarized below. We assume that the reader is already fairly familiar with it. Underlining denotes a vector or matrix, and, for typographical convenience, subscripts are often indicated in parentheses.

The system is assumed to have a large but finite number of states labeled i=1,2,...,S, corresponding to different molecular configurations. States occur according to the probability distribution vector $\underline{\pi}$, where the *ith.* element $\pi(i)$ is the normalized Boltzmann factor for state i. Transitions between states are made according to the stochastic transition matrix $\underline{P}$, which must satisfy the eigenvalue equation

$$\underline{\pi} = \underline{\pi P} \tag{1}$$

There are other conditions on $\underline{P}$, such as irreducibility (1), but we will not discuss them here. The matrix element p(i,j) is the probability of moving to state j in the next step, given that the system currently is in state i.

The Metropolis choice of $\underline{P}$ is

$$\begin{aligned} p(i,j) &= q(i,j)\cdot\min[1,\ \pi(j)/\pi(i)] \qquad & i \neq j \\ &= 1 - \sum_{k\neq i} p(i,k) & i = j \end{aligned} \tag{2}$$

The elements q(i,j) of $\underline{Q}$ are defined by the way trial configurations j are generated. Usually one of the N molecules is selected at random, and each of its coordinates is given a perturbation uniformly on some fixed interval $[-\Delta,\Delta]$ (for a spherically symmetric molecule). Thus the state j lies in some domain D(i) in configuration space centered on state i, with volume $V(D) = N(2\Delta)^3$. More concisely,

$$\begin{aligned} q(i,j) &= 1/V(D) \qquad & j \subset D(i) \\ &= 0 & \text{otherwise} \end{aligned} \tag{3}$$

Other relevant features of $\underline{Q}$ are

$$q(i,j) \geqslant 0 \tag{4}$$

$$\sum_j q(i,j) = 1 \tag{5}$$

$$q(i,j) = q(j,i) \tag{6}$$

The trial step generated by $\underline{Q}$ is accepted with probability $\min[1,\pi(j)/\pi(i)]$. A variation on this scheme, usually attributed to A. A. Barker (7), is to use a step acceptance probability of $1/[1 + \pi(i)/\pi(j)]$. It has been noted (3) that this algorithm actually was used earlier by Flinn and McManus (8).

To compute ensemble average properties (e.g., the mean potential energy) the relevant mechanical variable is formally represented by a vector $\underline{f}$, where $f(i)$ is the value of the variable in state i. A Markov chain of M states $\{X(m)\}$, $m=1,2,\ldots M$, is then generated using $\underline{P}$, and the average of f is estimated to be

$$\bar{f} = \frac{1}{M}\sum_{m=1}^{M} f[X(m)] \qquad (7)$$

The Hastings Formalism. Hastings (9) has published a generalization of the Metropolis algorithm. The essential features are that $\underline{Q}$ no longer need be symmetric as in eq. (6), and that any of a whole class of matrix functions $\underline{\alpha}$ may be used to determine step acceptance probabilities. More specifically,

$$\begin{aligned} p(i,j) &= q(i,j)\alpha(i,j) && i \neq j \\ &= 1 - \sum_{k\neq i} p(i,k) && i = j \end{aligned} \qquad (8)$$

We will not describe the allowed functional forms for $\underline{\alpha}$ except to note that the Metropolis and Barker algorithms, generalized for nonsymmetric $\underline{Q}$, represent two special cases:

$$\alpha_M(i,j) = \min[1,t(j,i)] \qquad (9)$$

$$\alpha_B(i,j) = 1/[1 + t(i,j)] \qquad (10)$$

where

$$t(i,j) = [\pi(i)q(i,j)/(\pi(j)q(j,i))] \qquad (11)$$

Hastings (and Metropolis *et al.*) considered only $\underline{P}$ which were reversible, i.e.,

$$\pi(i)p(i,j) = \pi(j)p(j,i) \qquad (12)$$

There exist non-reversible $\underline{P}$ which still satisfy eq. (1), but reversible $\underline{P}$'s generally are simpler to construct and deal with mathematically.

The Metropolis and (especially) the Hastings formalisms in principle give the MC practitioner wide latitude in choosing $\underline{P}$ for a particular problem. We now turn to the question of computational efficiency. In other words, how should $\underline{\alpha}$ and $\underline{Q}$ be chosen to produce a low error (i.e., variance) in $\bar{f}$ for a given investment in computer resources? In addition to empirical and intuitive rules, there now is some useful theoretical guidance on this question.

Optimization of Step Acceptance Matrix α. It has been widely conjectured that the Metropolis $\underline{\alpha}$ is superior to the Barker $\underline{\alpha}$, since $\alpha_M(i,j) \geq \alpha_B(i,j)$ when $i \neq j$ if the same $\underline{Q}$ is used for both. The Metropolis scheme will reject fewer trial steps; it will sample more states and presumably lead to smaller statistical errors. Peskun (10) has proved that this is so, and he has also obtained some much stronger results.

In this mathematical work, optimization refers to minimizing the asymptotic variance of $\bar{f}$, using $\underline{P}$ to sample from $\underline{\pi}$, where the asymptotic variance is defined as

$$v(\underline{f},\underline{\pi},\underline{P}) = \lim_{M\to\infty} \{M\cdot \mathrm{var}\ [\sum_{m=1}^{M} f(X(m))/M]\} \qquad (13)$$

The basic theorem is as follows. For two choices of $\underline{P}$, $\underline{P}_1$ and $\underline{P}_2$, assume that $\underline{P}_1 \geq \underline{P}_2$ in the sense that $p_1(i,j) \geq p_2(i,j)$ for all (i,j) where $i \neq j$. That is, all transitions to different states are at least likely for $\underline{P}_1$ as for $\underline{P}_2$. Then $v(\underline{f},\underline{\pi},\underline{P}_1) \leq v(\underline{f},\underline{\pi},\underline{P}_2)$.

It is clear that $\underline{P}_M \geq \underline{P}_B$ and, hence, that $\underline{P}_B$ gives an asymptotic variance which is at best as low as that for $\underline{P}_M$. For a discussion of the special case of sampling in a two-state system, see (11). Peskun also showed that $\underline{P}_M \geq \underline{P}$ for any other choice of $\underline{\alpha}$ within the Hastings formalism, so that the Metropolis scheme is asymptotically the optimum choice of $\underline{\alpha}$.

Optimization of Transition Matrix Q. In another paper (12), Peskun has investigated how the choice of $\underline{Q}$ affects the asymptotic variance, again within the Hastings formalism. The optimum $\underline{Q}$, which may not be unique (13), is asymmetric and reversible:

$$\pi(i)q(i,j) = \pi(j)q(j,i) \qquad (14)$$

If $\underline{Q}$ is reversible, then $\alpha_M(i,j) = 1$ for all (i,j), and $\underline{P} = \underline{Q}$. In other words, it is asymptotically optimum to place the whole burden of the sampling on $\underline{Q}$ so that no trial steps are rejected.

Eq. (14) does not give sufficient constraints to fix $\underline{Q}$. Peskun has shown further that the $\underline{Q}$ which minimizes $v(\underline{f},\underline{\pi},\underline{P})$ is a complicated function both of $\underline{f}$ and $\underline{\pi}$. Even though a rigorous optimization is impractical, by making a simplifying approximation he was able to come up with qualitative guidelines for choosing $\underline{Q}$ to reduce $v(\underline{f},\underline{\pi},\underline{P})$. The prescription is to relate $[q(i,j) - \pi(j)]$ roughly linearly to

$$[f(i) - f(j)]^2\ \pi(i) \qquad \pi(i) \geq \pi(j) \qquad (15)$$

$$\text{or} \qquad [f(i) - f(j)]^2\ \pi(j)^2/\pi(i) \qquad \pi(i) < \pi(j)$$

At the same time, Q should approximately satisfy eq. (14) and must rigorously satisfy eq. (4) and eq. (5). The prescription implies that transitions should be encouraged to states of high π and to states with values of f which differ greatly from that of f in the current state.

Next we will discuss how the usual choice of Q can be analyzed in terms of the guidelines. If Q is chosen according to eq. (3), f is ignored completely. The current state i of the system probably will have high π, as will nearby states j. "Nearby" means states with similar molecular coordinates. We wish to sample these states more frequently than distant states, about which we have no information except that most of them have $\pi \sim 0$ (in a dense fluid). A simple procedure is to set $q(i,j) = 0$ except for states lying in D(i), where D(i) corresponds to the perturbation of the coordinates of one or more molecules at state i. V(D) should be made as large as possible, consistent with including a high proportion of states with high π. In practice, V(D) is adjusted empirically by increasing Δ until the step acceptance fraction drops to ~0.5. Thus, there is some optimization of Q with respect to π.

Within D(i), q(i,j) should be made reversible. Unfortunately, this requires extensive knowledge of the behavior of π in D(i), which is computationally expensive to obtain. A zero-order approximation is that π is constant in D(i); this leads to the usual symmetric choice $q(i,j) = q(j,i)$. It often is not too expensive to obtain spatial analytical derivatives of potential energy functions, as long as the energy is being calculated anyway. One can make the first-order approximation that the gradient of the energy is constant in D(i) and is equal to its value at state i, then construct q(i,j) to be reversible under this assumption. Pangali *et al*. (14 and in this Symposium) have used essentially this technique to improve the rate of convergence of MC calculations on liquid water. A related method was used earlier for a quantum fluid, but with less success (15). The extension to higher derivatives of the energy is obvious.

Another approach is to improve Q by using information about the current configuration to select the size, shape, and position of D(i) at each step. For example, Rossky *et al*. (16) used the first derivative of the energy to select the origin of a Gaussian perturbation of the molecular coordinates.

In optimizing Q for a dense fluid, the importance of f usually will be less than that of π. The primary reason is that π is a strongly varying exponential function of the molecular coordinates, differing significantly from zero only in a small fraction of the states. The variation of most other functions (e.g., the potential energy) is weak by comparison. Since the true ensemble average of f is the sum of $f(i)\pi(i)$ over all states, states with high π dominate $\bar{f}$. Also, MC runs typically involve the calculation of several properties of the system; one $\underline{Q}$ generally will not optimize all of the estimates.

Nevertheless, optimizing $\underline{Q}$ with respect to $\bar{f}$ sometimes can be important. For example, the Helmholtz free energy difference A(1)-A(0) between two systems 1 and 0 can be written as

$$A(1) - A(0) = -kT\cdot\ln\langle\exp[-(U(1)-U(0))/kT]\rangle_0 \qquad (16)$$

where U(1) and U(0) are the potential energies and $\langle\rangle_0$ is the ensemble average in the 0 system. Here f, which is the exponential of the energy difference, may be as strongly varying as π, where $\pi \propto \exp[-U(0)/kT]$. Configurations where f is large but π is small may dominate $\bar{f}$. The umbrella sampling technique (2, 17) seeks to alter the distribution sampled from, in order to give more equal weight to states with large f but small π and those with small f but large π. Though the approach is different from that of Peskun, similarities do exist in the ways in which $\underline{f}$ enters into the construction of the transition matrices.

$\underline{f}$ is also important in the optimization of $\underline{Q}$ when $\underline{f}$ is constant over large classes of states. An example is in dilute solutions, where the most important solution properties are determined largely by solute-solvent and solvent-solvent interactions within the immediate solvation shell of the solute, and are relatively insensitive to the configurations of distant solvent molecules. Owicki and Scheraga (5) proposed an algorithm in which solvent molecules near the solute are perturbed more often than are distant ones. This was applied to calculations of the properties of dilute aqueous solutions of methane (18) and to calculations of the solvation free energy of hard spheres in the Lennard-Jones fluid (19). Part II of this chapter discusses a generalization of this algorithm. Squire and Hoover (20) used a similar but simpler procedure to calculate the free energy of vacancy formation in a Lennard-Jones crystal. In terms of the Peskun prescription, sampling preferentially near the solute makes q(i,j) large for transitions which are likely to involve large values of $[f(i)-f(j)]^2$, where $\underline{f}$ is a property sensitive to intermolecular interactions within the solvation shell.

How Useful are Peskun's Results? One obvious practical drawback is that a sampling algorithm with low asymptotic variance but high computational cost per step may be less efficient overall than another algorithm with higher asymptotic variance but lower cost per step. In fact, conflict of this type may well be common. Another drawback is that optimizing the asymptotic variance does not necessarily optimize the variance in an actual calculation, which consists of a finite number of steps. Whether the typical length of $\sim 10^6$ steps produces variances which behave asymptotically depends on factors such as the presence of quasi-ergodic problems (11).

Nevertheless, Peskun's results seem to be in gratifying agreement with intuition and with empirical observations. Prescription (12) is the best (and only?) mathematical guidance yet published for the optimization of $\underline{Q}$, even though it is hard to apply quantitatively. It is particularly important because it makes an estimate of the relative significance of $\underline{f}$ and $\underline{\pi}$ in the optimization. Clearly, this field is fertile ground for further innovations by mathematicians and MC practitioners alike.

II. An Algorithm for Sampling Preferentially Near the Solute

As was mentioned above, if one is interested in the structure and energetics of the solvation sphere about one solute molecule dissolved in N solvent molecules, it should be efficient to concentrate the sampling predominantly on those solvent molecules which are in the solvation sphere. A general algorithm based on this idea is outlined below.

Description of the Algorithm. Assume that a weighting function $\omega(r)$ has been defined, where r is the distance between a solvent molecule and the solute. The function ω may be chosen fairly arbitrarily. It need not be continuous, but it should be fast to evaluate, nonnegative, and a decreasing function of r. If the system is in state i, the next molecule to be perturbed is selected from the probability distribution $\underline{W}(i)=(W_1(i),W_2(i), \ldots W_N(i))$, where the components are defined by

$$W_K(i) = \omega(r_K) / \sum_{L=1}^{N} \omega(r_L) \tag{17}$$

and the r_L are the solute-solvent distances in state i. Obviously, $\underline{W}(i)$ is normalized. Once selected, the molecule is perturbed in the usual fashion.

The sampling from $\underline{W}(i)$ can be implemented in several ways. For example, a rejection sampling algorithm is outlined in Figure 1. An alternative is direct sampling: a random number ξ is chosen uniformly on [0,1], and molecule K is selected, where

$$\sum_{L=0}^{K-1} W_L(i) < \xi \leqslant \sum_{L=0}^{K} W_L(i) \tag{18}$$

and $W_0(i) \equiv 0$. Which algorithm is more efficient depends on N and the degree to which $\underline{W}(i)$ differs from a uniform distribution.

Selecting a molecule from $\underline{W}(i)$ makes $\underline{Q}$ asymmetric, because in general $\underline{W_K}(i) \neq \underline{W_K}(j)$. The proper Metropolis step acceptance probability is

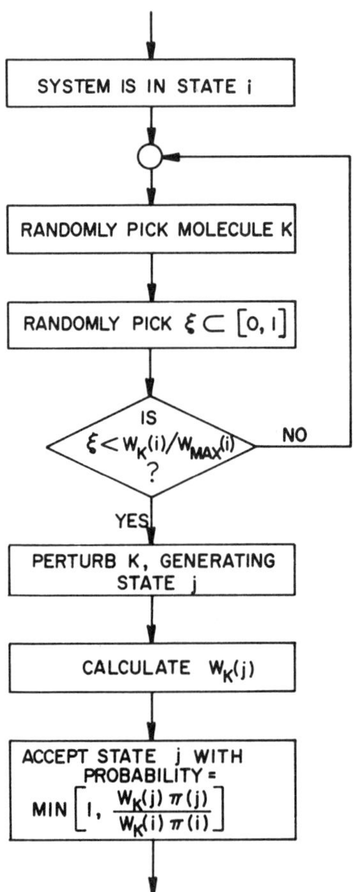

Figure 1. Rejection sampling algorithm for sampling from W(i). W_{MAX}(i) *is the largest element of* W(i).

$$\min[1,\ W_K(j)\ \pi(j)/(W_K(i)\pi(i))] \tag{19}$$

Example: Computation of the Solute-Solvent Radial Distribution Function g(r). The following calculation was done to test the effectiveness of a preferential sampling algorithm for reducing the statistical errors in the computed values of g(r) in a (two dimensional) hard disc solution for small values of r, where g(r) differs significantly from unity. The solution was constructed from N = 48 solvent particles with diameter σ and one solute particle with diameter 1.5σ. Square periodic boundary conditions were used, with cell side length 8.617σ. This corresponds approximately to 0.57 times the close-packing density of the solvent. With the maximum coordinate perturbation $\Delta = 0.2\sigma$, about 50% of the trial steps were accepted. The solute was perturbed every tenth step. An initial regular configuration was "equilibrated" to serve as the starting configuration for each of four computations of g(r), which were made using different sampling algorithms. The computed g(r), averaged over the four runs, is given in Figure 2. The length of each run was 300,000 steps.

In the first run, the solvent particles were sampled uniformly; in the second, a weighting function $\omega(r) = 1/r^2$ and the rejection technique in Figure 1 were used to sample preferentially near the solute. Figure 3 shows the resulting distribution W for a typical configuration. The calculated standard errors (21) of the calculated values of g(r) are given in Figure 4. For $r \lesssim 2.3\sigma$, the preferential sampling results (denoted by +) are more precise than those obtained with uniform sampling of the solvent particles (⊡). In this region, the average of the ratios of the errors is ~0.8. If the errors are behaving asymptotically, equal errors will obtain if roughly $(0.8)^2 \sim 0.6$ times as many steps are made in the preferential sampling run as in the uniform sampling run. Although there is a lot of scatter in the error data, we feel that this is a reasonable estimate of the efficiency increase.

For $r \gtrsim 2.3\sigma$, there appears to be no clear-cut difference between the two sets of errors. One would expect uniform sampling to give somewhat higher precision above $r \sim 2.7\sigma$ where, our data show, W typically drops below 1/N. Preferential sampling decreased the fraction of steps accepted only slightly, from 0.50 to 0.48.

In most molecular systems, the added costs of implementing preferential sampling are a small fraction of the total. If $\overline{W}_{MAX}$ is the ensemble-average largest value of W in a configuration, the mean number of trials to select a molecule by the rejection sampling algorithm in Figure 1 is $\overline{W}_{MAX} \cdot N$. In the calculation reported here, $\overline{W}_{MAX} \cdot N$ was found to be ~5. The extra

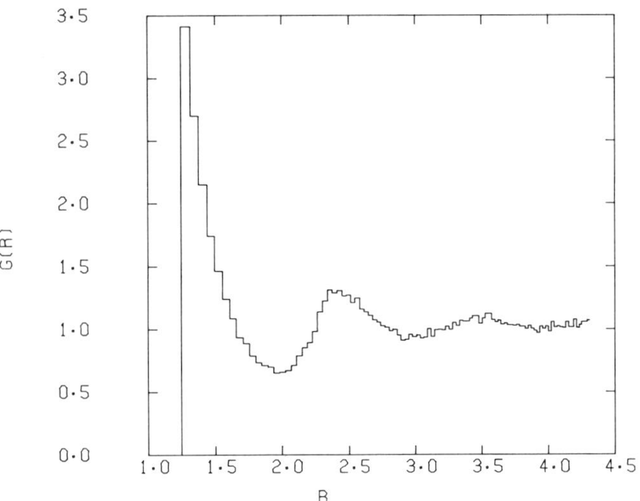

Figure 2. Computed solute–solvent radial distribution function g(r) *in hard disc system. Unit distance is solvent diameter* σ. *Data points were tabulated over 100 intervals, equally spaced in* r^2.

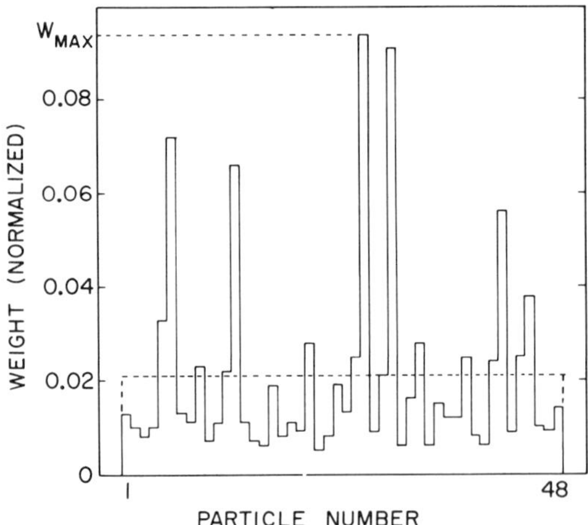

Figure 3. Probability distribution function W *for a typical hard disc configuration, with* $\omega(r) = 1/r^2$. W_{MAX} *is the largest element, and the dashed line near* W = *0.02 represents a uniform distribution,* $W_L = 1/N$ *for each particle.*

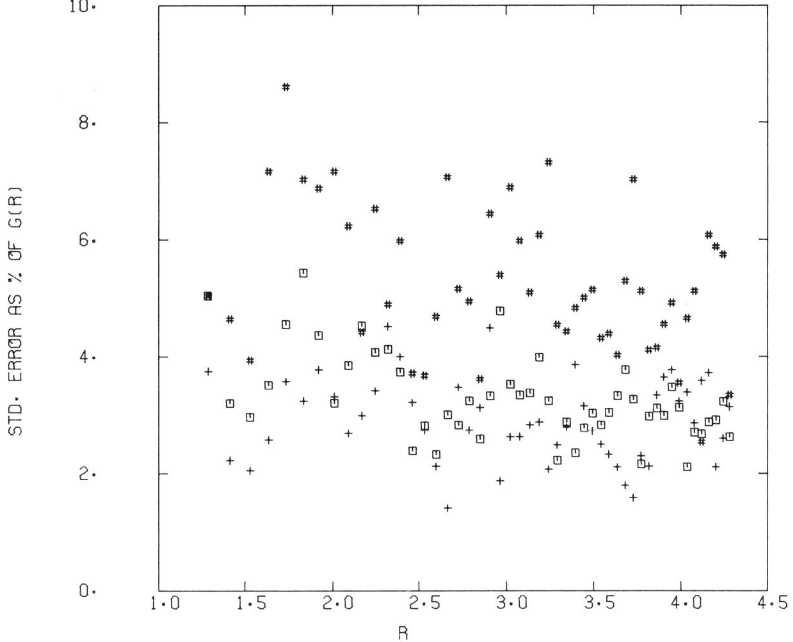

Figure 4. Comparison of standard errors for g(r), *when* g(r) *is computed with different sampling algorithms. (□) uniform sampling of solvent particles, solute perturbed every tenth step; (+) preferential sampling of solvent particles, solute perturbed as above; (#) uniform sampling of solvent particles, solute fixed. For clarity, only every other data point is plotted. Estimates of errors are based on fluctuations in* g(r) *between blocks of 20,000 configurations (21).*

computing costs, compared with uniform sampling, were about 20 seconds of IBM 370/168 CPU time per 10^5 steps.

Digression: How Important Is It to Move the Solute? Perturbing a solvent molecule generates only one new solute-solvent distance for calculating g(r), while moving the solute generates N new distances. This suggests that one should move the solute more often than a solvent molecule, on the average. Note that it is not necessary to move the solute at all, since only relative molecular coordinates are important and thus the coordinate system can be fixed on a molecule (e.g., the solute). To test the importance of this effect, the two computations discussed above were repeated, except that the solute was not perturbed. The standard errors for the uniform sampling run are plotted (#) in Figure 2. They usually lie well above the errors from both runs in which the solute moved. For clarity, the solute-fixed/preferential sampling results are not plotted. They are not too different from the solute-moved/preferential sampling results for small r, but are significantly higher for large r. Not surprisinly, it is important to perturb the solute.

Acknowledgements

Thanks go to P. Peskun for a copy of his paper (12) as well as for interesting discussions of the relevance of his results to calculations on dense fluids. The author is supported by an Institutional Research Fellowship under NIH Grant No. GM 07026.

Literature Cited

1. Hammersley, J. M., and Hanscomb, D. D., "Monte Carlo Methods," especially Chapter 9, Methuen, London, 1964.
2. Valleau, J. P., and Torrie, G. M., Chapter 5 in Berne, B. J. (ed.), "Statistical Mechanics, Part A," Plenum, New York, 1977.
3. Wood, W. W., and Erpenbeck, J. J., Ann. Rev. Phys. Chem., (1976) 27, 319.
4. Bennett, C. H., J. Comp. Phys., (1976), 22, 245.
5. Owicki, J. C., and Scheraga, H. A., Chem. Phys. Lett., (1977), 47, 600.
6. Metropolis, N., Rosenbluth, A. W., Rosenbluth, M. N., Teller, A.H., and Teller, E., J. Chem. Phys., (1953), 21, 1087.
7. Barker, A. A., Austral. J. Phys., (1965), 18, 119.
8. Flinn, P. A., and McManus, G. M., Phys. Rev., (1961), 124, 54.
9. Hastings, W. K., Biometrika, (1970), 57, 1.
10. Peskun, P. H., Biometrika, (1973), 60, 3.
11. Valleau, J. P., and Whittington, S. G., Chapter 4 in Berne, B. J. (ed.), "Statistical Mechanics, Part A," Plenun, New York, 1977.

12. Peskun, P. H., "Guidelines for Choosing the Transition Matrix in Monte Carlo Sampling Methods Using Markov Chains," presented at the Fourth Conference on Stochastic Processes and Their Applications, York Univ., Toronto, August 5-9, 1974. For an abstract, see Peskun, P. H., Adv. Appl. Probab., (1975), 7, 261.
13. Peskun, P. H., private communication.
14. Pangali, C., Rao, M., and Berne, B. J., Chem. Phys. Lett., in press.
15. Cepereley, D., Chester, G. V., and Kalos, M., Phys. Rev. B, (1977), 16, 3081.
16. Rossky, P. J., Doll, J. D., and Friedman, H. L., J. Chem. Phys., in press.
17. Torrie, G., and Valleau, J. P., J. Comp. Phys., (1977), 23, 187.
18. Owicki, J. C., and Scheraga, H. A., J. Am. Chem. Soc., (1977), 99, 7413.
19. Owicki, J. C., and Scheraga, H. A., J. Phys. Chem., (1978), 82, 1257.
20. Squire, D. R., and Hoover, W. G., J. Chem. Phys., (1969), 50, 701.
21. Wood, W. W., Chapter 5 in Temperley, H. N. V., Rowlinson, J. S., and Rushbrooke, G. S. (eds.), "Physics of Simple Liquids," North-Holland, Amsterdam, 1968.

RECEIVED September 7, 1978.

15

Molecular Dynamics Simulations of Simple Fluids with Three-Body Interactions Included

J. M. HAILE

Chemical Engineering Department, Clemson University, Clemson, SC 29631

A primary goal in the study of the fluid states of matter is an ability to predict all static and dynamic properties using only information about the constituent molecules. The fluid systems of interest here are those in which the temperature and molecular masses are sufficiently high that quantum effects are negligible and for which the classical Hamiltonian can be assumed separable into independent translational, rotational, internal, and configurational contributions. The principal remaining difficulty in realizing the goal of complete property prediction for such systems is quantitative description of the intermolecular forces within the fluid. These forces are commonly discussed in terms of an intermolecular potential energy U which, for a fluid of N spherical molecules, depends only on the positions of the molecular mass centers $\underline{r}_i$:

$$U = U(\underline{r}_1, \underline{r}_2, \ldots \underline{r}_N) \qquad (1)$$

This intermolecular potential is usually written as a series of terms which individually account for 2-body, 3-body,... N-body interactions:

$$U = \sum_{i<j}\sum u(\underline{r}_i\underline{r}_j) + \sum_{i<j<k}\sum\sum u(\underline{r}_i\underline{r}_j\underline{r}_k) \qquad (2)$$

$$+ \sum_{i<j<k<\ell}\sum\sum\sum u(\underline{r}_i\underline{r}_j\underline{r}_k\underline{r}_\ell) + \ldots$$

Much progress in recent years has been made towards understanding the fluid state via the "pair theory" of fluids wherein the series in eq. (2) is truncated after the first term (see, for example, reference (1)), i.e.,

$$U \simeq \sum_{i<j}\sum u(\underline{r}_i\underline{r}_j) \qquad (3)$$

Even the simplest real fluids (the inert gases) do not obey the pairwise additive assumption eq. (3) at moderate to high densities (2). However, the advent of the Monte Carlo (3) and

0-8412-0463-2/78/47-086-172$05.00/0

molecular dynamics (4) techniques for simulating fluids on computers has enabled generation of a vast amount of data for model fluids which obey eq. (3) exactly. When compared with various theoretical calculations (such as from integral equations, perturbation theories, etc.) the simulation results provide tests of the theories which are consistent in the assumption of pairwise additivity. When compared with experimental measurements of the properties of real fluids, the simulation results can be used to directly test the adequacy of the pairwise additive assumption. The results obtained from these kinds of studies have enhanced our understanding of the relations between features of the intermolecular potential and fluid properties. This new knowledge has, in turn, prompted refinements in existing pair potential models and construction of new potential functions.

Although the computer simulation of fluids is now recognized as a powerful tool for testing models for the intermolecular potential, almost no simulation work has been reported in which, even, three body forces are directly taken into account. The deterrent to performing such simulations has been the excessively large amount of computer time required. This paper reports development of a method for performing molecular dynamics simulations in acceptable amounts of computer time for 108 spherical molecules with long range, three body interactions explicitly included. The technique reported here is a variant of the multiple time step methods recently developed by Streett and coworkers (5). A description of the multiple time step method applied to nonspherical molecules is given elsewhere in this volume. Of the sections which follow: Section 1 gives a brief summary of previous studies of nonadditivity effects: Section 2 contains a description of the triple dipole Axilrod-Teller potential used in this study; Section 3 presents the molecular dynamics method developed; Section 4 reports preliminary results for the three body effects on equilibrium properties.

1. Summary of Previous Work on Nonadditivity.

Over the years a number of attempts have been made to calculate nonadditive contributions to fluid and crystalline properties. Most of this work has been concerned with nonadditive three body contributions to equilibrium properties, particularly: virial coefficients (6, 7, 8, 9, 10), internal energy (11), free energy (12), and the radial distribution function (10, 13, 14,15). Additional references are cited by Barker, *et al.* (11).

In contrast to the large number of theoretical studies, extremely few computer simulation studies of nonadditivity effects have been performed. This dearth of simulation work is due to the prohibitively large amount of central processor time required to sample $N(N-1)(N-2)/3!$ triplets in a system of N particles. Barker and coworkers (11) have used both Monte Carlo and molecular dynamics to estimate three body contributions to the internal energy and pressure of liquid argon. However, the nonadditive contributions to the properties were obtained by a perturbative technique in which averages over the three body potential were evaluated from a subset of the entire simulation (every 1000th configuration in the case of Monte Carlo). Repeated, time-consuming sampling of all possible triplets in the fluid was thereby avoided. Recently, Schommers has performed molecular dynamics simulations of a two-dimensional Lennard-Jones fluid; the calculations were done in two dimensions rather than three because there are significantly fewer triplets of molecules in two dimensions (16).

As study of nonadditive contributions to properties has developed, uncertainty has arisen as to the relative importance of short and long range three body interactions. Long range interactions are usually taken to be in the form of a multipole expansion of dispersion forces (17) in which the Axilrod-Teller triple dipole potential (18, 19) is the leading term. The short range interactions are often considered to be those due to nonadditive charge overlap or exchange forces. In the simulation work by Barker, et al. the Axilrod-Teller potential was assumed to be the only significant nonadditive interaction (11, 12). The simulation using the Axilrod-Teller model with a true pair potential gave values of pressure and internal energy in good agreement with experimental values for liquid argon. On the other hand, Jansen and Lombardi argue that short range three body overlap interactions explain the stabilization of rare gas crystals in an fcc structure rather than the expected hcp structure (20). This question of the relative importance of short and long range nonadditive interactions apparently remains unresolved.

The magnitudes of three body contributions to fluid properties have been found to depend on the sensitivity of the property to the intermolecular potential. Thus, Barker, et al. found that the Axilrod-Teller potential contributes about 50% to the pressure but only about 5% to the internal energy in liquid argon (12). The effect of three body forces on the radial distribution function is generally considered to be small (10); although Singh, et al. claim significant three body effects on the radial distribution function based on thermodynamic perturbation theory calculations for a Lennard-Jones plus Axilrod-Teller fluid (14). Barker has obtained the radial distribution function for liquid argon from a simulation using the Barker-Fisher-Watts (BFW) pair potential plus the Axilrod-Teller model and quantum corrections (13, 21). This radial distribution function agrees very well with that obtained from neutron diffraction experiments by Yarnell, et al. (13). However, the Lennard-Jones radial distribution function determined by Verlet (22) using molecular dynamics agrees equally well with the neutron scattering results (13, 21). The conclusion would seem to be that the radial distribution function is not very sensitive to the details of the potential and, especially at high density, is largely determined by short range, repulsive forces. Singh, et al. argue, however, that the BFW and Lennard-Jones are very different pair potentials and, therefore, their radial distribution functions must be very different. Hence, they expect the three body contribution to the radial distribution function to be large (14). Schommers' two-dimensional Lennard-Jones plus Axilrod-Teller simulations show small effects of the three body interaction on the radial distribution function (16).

Little is known concerning three body contributions to dynamic fluid properties. Fisher and Watts have calculated the self diffusion coefficient using the BFW pair potential without including the Axilrod-Teller interaction and obtained results similar to those for a Lennard-Jones fluid at the same densities (23). Schommers' two dimensional Lennard-Jones plus Axilrod Teller simulations show significant three body effects on the velocity autocorrelation function; however, the two dimensional self diffusion coefficient is little affected (16). Schommers is careful to point out, however, that results found in two-dimensional fluids do not necessarily extrapolate to three dimensions.

Little study has been made of nonadditive effects in fluids composed of nonspherical molecules. Singh and coworkers have developed a thermodynamic perturbation theory for this purpose (24, 25). Stogryn has given the analog of the Axilrod-Teller potential for nonspherical molecules (26).

2. Potential Model Used

The potential model used to test the molecular dynamics method described below was a Lennard-Jones 6, 12 pair potential plus Axilrod-Teller triplet potential. The Lennard-Jones model is given by:

$$u_{LJ}(r_{ij}) = 4\varepsilon[(\sigma/r_{ij})^{12} - (\sigma/r_{ij})^{6}] \tag{4}$$

while the Axilrod-Teller model is given by (17, 18, 19):

$$u_{AT}(r_{ij},r_{ik},r_{jk}) = \frac{\nu[1 + 3\cos\theta_1\cos\theta_2\cos\theta_3]}{(r_{ij}r_{ik}r_{jk})^3} \tag{5}$$

where r_{ij} is the scalar distance between atoms i and j, etc., θ_i is the angle formed at atom i by r_{ij} and r_{ik}, etc., and ν is a species dependent constant. The value used for ν in this work is that quoted by Barker, et al. (12) for argon: $\nu = 73.2(10^{-84})$ erg cm^9. Using the usual values for the Lennard-Jones pair potential parameters, namely $\varepsilon/k = 120.\ ^{\circ}K$, $\sigma = 3.405$ Å, gives $\nu^* \equiv \nu/\varepsilon\sigma^9 = 0.0718$.

Figures 1 and 2 show the Axilrod-Teller potential for several representative triangular shapes formed by three atoms. These figures indicate that the Axilrod-Teller potential is

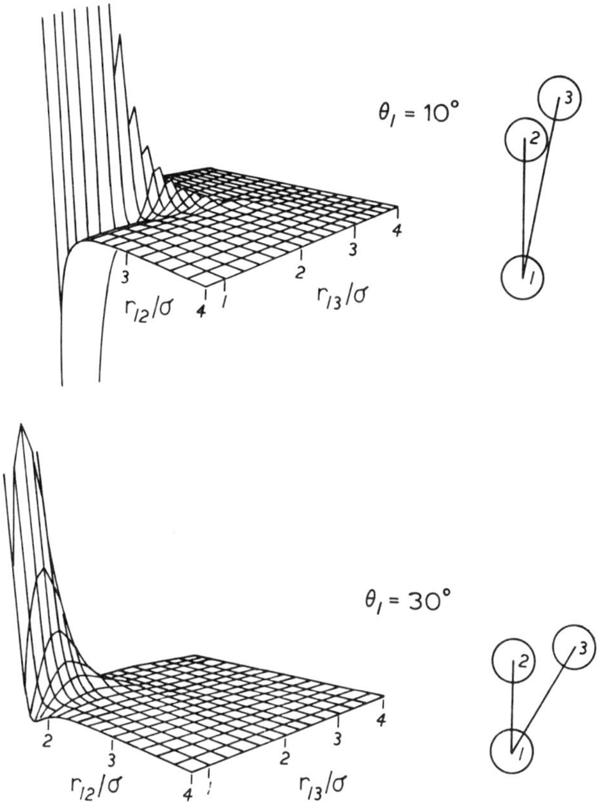

Figure 1. Axilrod–Teller triple dipole potential for representative triangles formed by three atoms. The angle θ_1 is that between r_{12} and r_{13}.

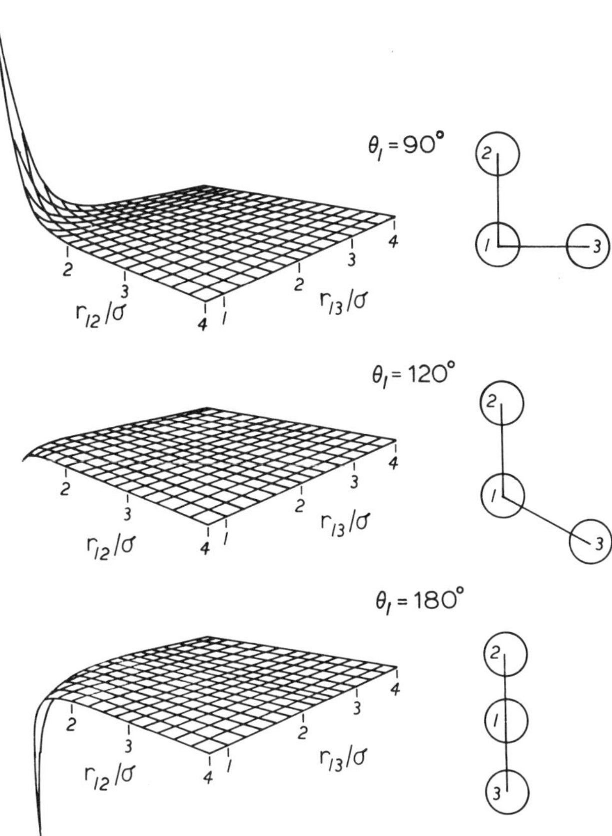

Figure 2. Axilrod–Teller triple dipole potential for representative triangles formed by three atoms. The angle θ_1 is same as in Figure 1.

repulsive over most acute triangles and attractive over most obtuse triangles. The repulsive parts of the interaction appear to be stronger and of longer range than the attractive portions. In the simulations reported here the Lennard-Jones potential was truncated at $r_c = 2.5\sigma$. Figures 1 and 2 indicate that the Axilrod-Teller potential is of shorter range than the Lennard-Jones model for most triangular shapes; hence, the Axilrod-Teller potential was truncated when any one of the lengths r_{ij}, r_{ik}, or r_{jk} was greater than $r_c = 2.5\sigma$:

$$u_{AT}(r_{ij}, r_{ik}, r_{jk}) = 0 \quad \text{if } r_{ij}, r_{ik}, \text{ or } r_{jk} > r_c \tag{6}$$

A similar truncation of this potential was used by Barker, *et al.* (12).

3. Molecular Dynamics with Three-Body Interactions

3.1 Conventional Molecular Dynamics (CMD). When three body interactions are explicitly included in molecular dynamics simulations, the equations of motion to be solved take the form:

$$\underline{F}_i = m_i \frac{d^2\underline{r}_i}{dt^2} \tag{7}$$

$$\underline{F}_i = \underline{F}_i^{2B} + \underline{F}_i^{3B} \tag{8}$$

$$\underline{F}_i^{2B} = - \sum_{i \neq j} \frac{\partial u^{2B}(r_{ij})}{\partial \underline{r}_i} \tag{9}$$

$$\underline{F}_i^{3B} = - \sum_{j<k\neq i}\sum \frac{\partial u^{3B}(r_{ij}, r_{ik}, r_{jk})}{\partial \underline{r}_i} \tag{10}$$

where $\underline{F}_i$, m_i, $\underline{r}_i$ are the force, mass and position vector of mass center for particle i, respectively, and superscripts 2B and 3B denote two and three body contributions, respectively. In the work reported here these equations of motion were solved using a predictor-corrector scheme due to Gear (27). Details of application of this method to molecular dynamics simulations are given elsewhere (28, 29). The geometry of the system was taken to be a cube of side L and periodic boundary conditions were used with the usual minimum image criterion (30).

In molecular dynamics with only pair interactions, the speed of program execution is limited by evaluation of the force on each particle, eq. (9). Since simulations usually utilize pair potentials truncated at some pair separation $r_c \leq L/2$, a significant increase in program execution speed can be obtained by maintaining some type of list of particles in order to avoid sampling those particles separated by distances greater than r_c. At least two types of such lists are in use. One is a neighbor list (31) maintained for each particle i which tabulates all particles that are within a distance r_L of the particle i. The list distance r_L is chosen to be slightly longer than the cutoff distance r_c. The second type is a cell list (32, 33) in which the system is divided into small cells and a list is maintained of the particles within each cell. In this method, pair interactions are sampled only between particles in adjacent cells. In both methods the contents of the lists must be updated at periodic intervals throughout the simulation.

When three body interactions are included in the simulation, the speed limiting step becomes evaluation of the three body forces of eq. (10). Now, if the truncation of the three body potential is taken to be that given by eq. (6) with r_c having the same value for both the two and three body potentials, then the _same_ list used to speed formation of the sum in eq. (9) can be applied to the double sum in eq. (10); i.e., no increase in storage is required. For example, if a neighbor list is used, then the double sum in eq. (10) is formed only over those molecules held in the list. Forming triplets using molecules from the list with any molecule not in the list automatically gives:

$$r_{ij}, r_{ik}, \text{ or } r_{jk} > r_L > r_c \tag{11}$$

and the truncation of eq. (6) applies. This algorithm in which a neighbor list is used for both two and three body force evaluation and in which three body forces are calculated explicitly at each time step shall be referred to as the conventional molecular dynamics (CMD) method.

Unfortunately, even with the neighbor list applied to the three body forces, the CMD program for 108 Lennard-Jones plus Axilrod-Teller particles executes about 15 times slower than the same program for 108 Lennard-Jones particles. In the work reported here, source programs were written in FORTRAN IV using single precision arithmetic, compiled on an IBM FORTRAN IV-G compiler, and executed on the IBM 370/165 at Clemson University.

3.2 Multiple Time Step (MTS) Method Applied to Three Body Forces. In order to improve the execution speed of simulation programs with three body interactions included, the multiple time step method of Streett, _et al._ (5) has been applied to the evaluation of the three body forces. The multiple time step method attempts to take advantage of the fact that molecular motions in a fluid may be reduced to components which operate on very different time scales. More precisely, one can often identify components of the force on a molecule which have relatively large differences in their rates of change with time. It is the quickly varying component of the force which limits the size of the time step Δt which must be used to obtain stable solutions to the differential equations of motion. In the MTS method these quickly varying components of the force are explicitly evaluated at each time step in the usual way; however, the slowly varying components are only explicitly evaluated at longer intervals, say, every n time steps. At time steps between explicit determination, the slowly varying components of the force are estimated by extrapolation from the previous explicit evaluation. The extrapolation may be linear (LMTS method) or, if one is more ambitious, based on a Taylor series expansion. The degree to which the MTS method improves execution speed of the program depends on the amount of computation avoided in extrapolating the slowly varying force components rather than explicitly calculating them.

If we consider applying this MTS method to simulations with three body interactions, eq. (8) already represents the force on molecule i divided into two parts. Barker and Henderson have noted that the three body Axilrod-Teller interaction is slowly varying (21). To test this, we performed a molecular dynamics simulation for 108 particles interacting with Lennard-Jones plus Axilrod-Teller potentials using the conventional molecular dynamics algorithm. Figure 3 shows a typical comparison of a component of the two and three body forces on one particle during a segment of the simulation. The figure indicates that the three body component of the force does indeed have a slower rate of change than the two body force component.

Encouraged by Figure 3, the MTS method was then applied to the simulation in the following manner. The entire three body force is considered to be slowly varying, therefore all of $\underline{F}_i^{3B}$ is explicitly evaluated only at periodic intervals in the simulation. We choose to use linear extrapolation of $\underline{F}_i^{3B}$ for, although Streett, et al. find that a third order Taylor series extrapolation is more accurate, the analytic evaluations of the time derivatives of the Axilrod-Teller force are hopelessly tedious. Thus, $\underline{F}_i^{3B}$ is explicitly evaluated at two successive times, t_o and $(t_o + \Delta t)$, and then linearly extrapolated over the next n time steps. This gives the three body force on particle i at any intermediate time step $(t_o + k\,\Delta t)$ as:

$$\underline{F}_i^{3B}(t_o + k\,\Delta t) = \underline{F}_i^{3B}(t_o) + k[\underline{F}_i^{3B}(t_o + \Delta t) - \underline{F}_i^{3B}(t_o)] \quad (12)$$

In a simulation program using a predictor-corrector algorithm, eq. (12) would appear in the force evaluation step. The only additional storage required for the MTS method over that for the CMD method is space for 6N numbers - 3N locations for the components of $\underline{F}_i^{3B}(t_o)$ and 3N locations for the components of the difference term in brackets in eq. (12).

We experimented with various values of the time step Δt and number of extrapolation steps n. A compromise among stable solutions of the differential equations, conservation of system energy, and execution speed was obtained using n = 8 and

$\Delta t = 2.15\ (10^{-15})$ sec. This value of the time step is about one-fifth that used by Verlet (31) and Rahman (34) in simulations of the pairwise additive Lennard-Jones fluid.

Note that, in the MTS method, any property which is being calculated as a running ensemble average at each time step during the simulation must also be divided into slowly (three body) and quickly (two body) varying contributions. The MTS method is applied to the slowly varying components of the properties via equations analogous to eq. (12). This comment applies, for example, to the internal energy and pressure in the work reported here.

Two parallel simulations for 108 particles interacting with Lennard-Jones plus Axilrod-Teller potentials have been performed. The first calculation utilized the CMD method in which the forces were explicitly evaluated at each time step. In the second run the two body forces were determined in the standard way and the LMTS method described above was applied to the three body forces. Both runs were started from the same initial particle positions and velocities and both were continued for 1650 time steps. A comparison of the properties obtained from the two calculations is given in Table I. In addition to the properties listed in Table I, radial distribution functions, velocity, speed, and force autocorrelation functions, and atomic mean squared displacements (from which diffusion coefficients were obtained) were calculated. For all of these properties, the LMTS values were within 0.1% of the values obtained by the CMD method. Figure 4 shows the per cent deviation in the instantaneous total energy of the two calculations.

Since, in the LMTS method described above, the three body forces are explicitly evaluated only two of every ten time steps, we might expect the LMTS method to be about 5 times as fast as the CMD method. The results in Table I confirm this; i.e., using the LMTS method, a simulation of 108 Lennard-Jones plus Axilrod-Teller particles requires about 20 CPU minutes for 2000 time steps on an IBM 370/165. This execution speed remains about 2.5 times slower than a CMD method, pairwise additive Lennard-Jones simulation.

The results from the LMTS method and the CMD method are in good agreement for those properties which are averaged over the simulation. However, after about 1000 time steps, the instantaneous components of the three body forces on individual particles in the LMTS calculation begin to deviate from those in the CMD method, as indicated in Table I. These deviations are due to gradual accumulation of inaccuracies from the linear extrapolation of the three body force. It should be emphasized that while these deviations occur, the total energy and momentum of the LMTS system remain conservative and agree closely with those of the CMD method (see Figure 4). This disagreement in the three body forces can be interpreted as a drifting of the

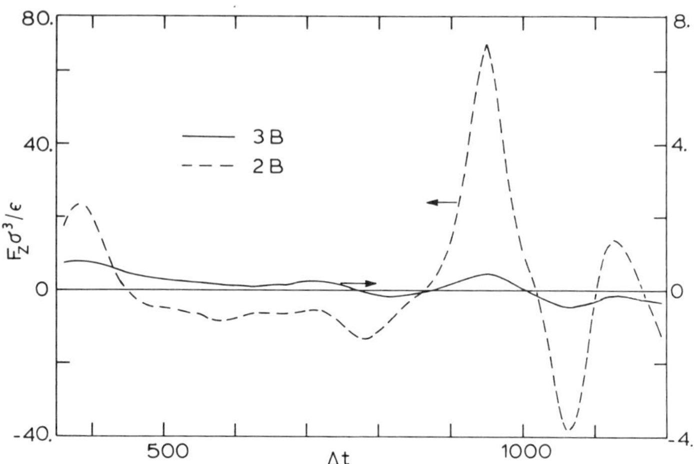

Figure 3. Comparison of z-component of two-(2B) and three-body (3B) forces on particle #27 in conventional molecular dynamics simulation of Lennard–Jones plus Axilrod–Teller fluid. Δt = *time step* = *2.15* (10^{-15}) *sec,* $\rho\sigma^3 = 0.65$, $kT\epsilon^{-1} = 1.033$. *Note that the scale for three-body force is an order of magnitude smaller than that for the two-body force.*

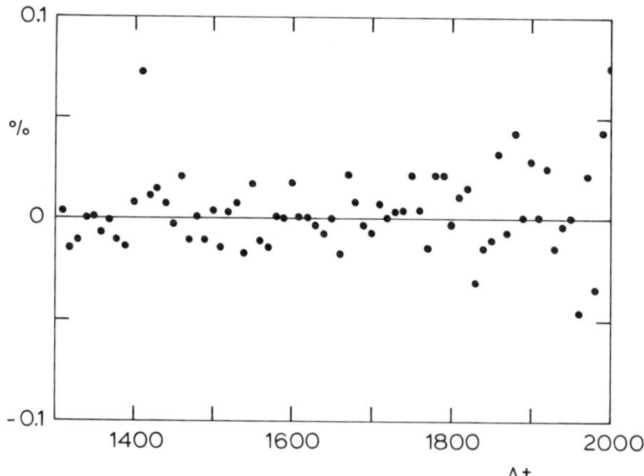

Figure 4. Percent deviation in instantaneous total system energy calculated by MTS method from that determined by CMD method for 108 particles interacting with Lennard–Jones plus Axilrod–Teller forces. System parameters same as those in Figure 3. Percent deviation calculated as: $(E_{CMD} - E_{MTS})100/E_{CMD}$.

Table I

Test of the Linear Multiple Time Step Method Applied to the Axilrod-Teller Potential in Molecular Dynamics Simulations of Lennard-Jones plus Axilrod-Teller Interactions.

		CMD Method	LMTS Method
Number Particles, N		108	108
Number density, $\rho\sigma^3$		0.65	0.65
Time step, Δt (10^{15})sec		2.15	2.15
CPU time, mins.		71.2	17.1
Temperature, $kT\varepsilon^{-1}$		1.0325	1.0326
Average Conf. Int. Energy, $U(N\varepsilon)^{-1}$		-4.356	-4.356
Average Pressure, $P(\rho kT)^{-1}$		0.104	0.103
Instantaneous Total Energy $E(N\varepsilon)^{-1}$ at time step 1650		-2.795	-2.797
Instantaneous Components of 3-body Force	$F_x^{3B}\sigma^3\varepsilon^{-1}$	-0.363	-0.373
on particle 27 at time	$F_y^{3B}\sigma^3\varepsilon^{-1}$	0.125	0.120
step 1650	$F_z^{3B}\sigma^3\varepsilon^{-1}$	0.0465	0.0519
Components of position	$x\sigma^{-1}$	1.457	1.462
vector for particle 27	$y\sigma^{-1}$	4.981	4.987
at time step 1650	$z\sigma^{-1}$	1.358	1.364

Table II

Preliminary Results Showing Effect of Three Body Axilrod-Teller Force on Internal Energy and Pressure

	$\rho\sigma^3 = 0.65$			$\rho\sigma^3 = 0.817$	
	Lennard-Jones		LJ + AT	Lennard-Jones	LJ + AT
	Verlet (ref. (31))	This Work	MTS Method	This Work	MTS Method
Number Particles	864	108	108	108	108
Time Step (10^{15}) sec.	9.6	2.15	2.15	2.15	2.15
Number Time Steps	1200	2000	1650	2000	1730
$kT\varepsilon^{-1}$	1.036	1.052	1.033	0.746	0.746
$U(N\varepsilon)^{-1}$	-4.52	-4.54	-4.36	-5.90	-5.64
$P(\rho kT)^{-1}$	-0.11	-0.05	+0.10	-0.20	+0.38

LMTS system onto a different trajectory in phase space from that followed by the system evolved by the CMD method. Hence, there is no strictly Newtonian connection between widely separated phase points; but then, there is no true Newtonian connection between widely separated phase points in the system generated by the CMD method due to round off errors and the truncated potential. The conclusion is that the phase space trajectories become different in the two calculations, but one is not necessarily more wrong (or right) than the other. Hoover and Ashurst have discussed this point in another context with similar conclusions (35). The restriction which this imposes is that the LMTS method cannot be used to study very long-lived phenomena; such as the long tail of the velocity autocorrelation function.

4. Preliminary Results for Equilibrium Properties

The linear multiple time step method described in Section 3 has been used to simulate 108 particles interacting with Lennard-Jones plus Axilrod-Teller potentials at two state conditions: (a) $\rho\sigma^3 = 0.65$, $kT\varepsilon^{-1} = 1.036$, which is one of the conditions of the Lennard-Jones fluid studied by Verlet (31) and by Singh, et al. (14, 15), (b) $\rho\sigma^3 = 0.817$, $kT\varepsilon^{-1} = 0.746$, which is close to the condition of the Lennard-Jones fluid studied by Rahman (34). From these two calculations the configurational internal energy, pressure, and radial distribution function have been determined. For comparison, simulation runs for a purely Lennard-Jones fluid have also been performed at state conditions close to the above.

Equilibrium property results at both state conditions for the Lennard-Jones and Lennard-Jones plus Axilrod-Teller fluids are compared in Table II. Long range corrections for the truncated two body potential have been added to the internal energy and pressure; however, long range corrections for the truncated three body potential have been neglected in the values given in Table II. These results confirm the earlier work by Barker, et al. in that the three body effect is only a few percent on the internal energy, but is significantly larger on the pressure. Note that, at the state conditions studied, the net effect of the Axilrod-Teller interaction is repulsive, as evidenced by positive contributions to the energy and pressure.

In Figure 5 the radial distribution functions for the Lennard-Jones and Lennard-Jones plus Axilrod-Teller fluids are compared at the high density state condition. It appears that the three body interactions have no effect except in the first peak region where the triplet interactions lower g(r) by about 3%. However, we estimate the statistical error in g(r) for these simulations is about ± 3%, so it is unclear to what extent the observed effect is real. We note that Schommers has reported small three body effects on the distribution function for the two dimensional fluid, as well (16). However, the perturbation theory calcula-

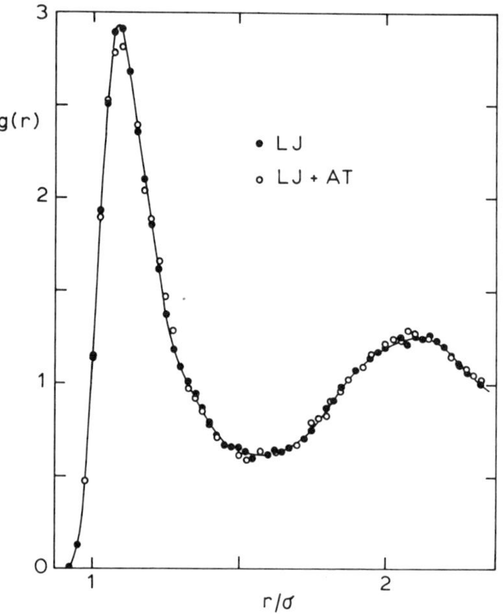

Figure 5. Effect of Axilrod–Teller potential on radial distribution function at $\rho\sigma^3 = 0.817$. For Lennard–Jones fluid, $kT\epsilon^{-1} = 0.746$; for Lennard–Jones plus Axilrod–Teller fluid, $kT\epsilon^{-1} = 0.740$.

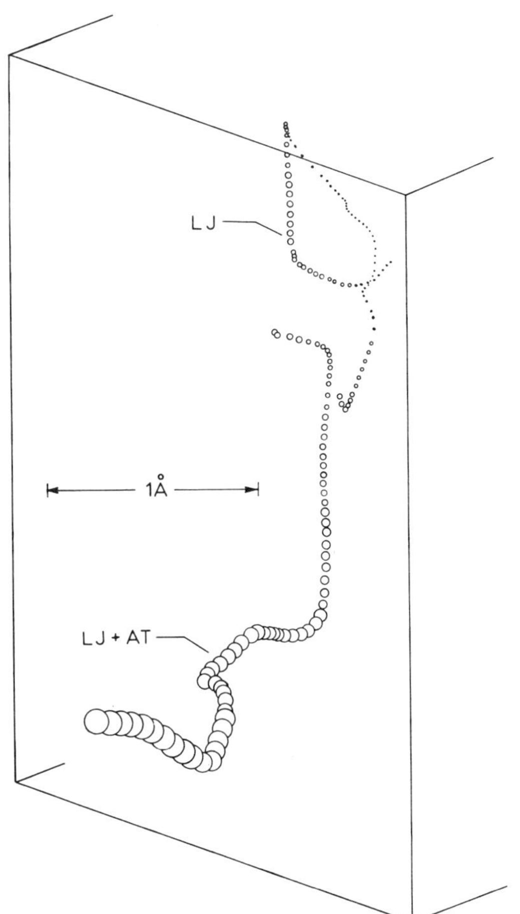

Figure 6. Comparison of trajectories of particle #27 in Lennard–Jones and Lennard–Jones plus Axilrod–Teller fluids at $\rho\sigma^3 = 0.65$. Both runs were started from the same point in phase space and trajectories shown are from time steps 500–2000 in each simulation. For this calculation the Axilrod–Teller strength constant ν was assigned a value of three times that for argon given in Section 2. Note that the circles represent positions of the center of mass of the atom, not the atomic diameter.

tions of Singh and Ram show "significant" effect of the triplet interactions on g(r) at low temperatures and high densities (14). The g(r) for the low density calculation shows essentially no effect due to the three body forces. Singh and Ram also find little effect at low densities.

Although some dynamic properties have been calculated from these simulations, we find, in general, that 2000 time steps for 108 particles do not provide sufficient data to obtain statistically meaningful dynamic properties. Work is now in progress aimed at increasing the number of particles in the system and the length of the run to improve the statistical precision of dynamic property calculations.

In an effort to gain further enlightenment as to the effects of the Axilrod-Teller potential, we have compared individual particle trajectories for the same particle over the same time segment from two different simulations. In one simulation the intermolecular potential was Lennard-Jones while in the other, the potential was Lennard-Jones plus Axilrod-Teller. Both calculations were done at the same density and were started from the same configuration and particle velocities. Figure 6 compares the trajectories obtained from such a study. In the Lennard-Jones fluid the particle shown in Figure 6 is exhibiting vibratory motion due to collisions with neighboring particles. In the Lennard-Jones plus Axilrod-Teller fluid, the same particle during the same time segment is undergoing an extended trajectory of largely diffusional motion. It may be that study of three body forces on dynamic properties will reveal fairly strong effects on individual particle motion but that these effects tend to cancel when averaged to obtain bulk dynamic properties. Hence, Schommers' two dimensional simulations show strong three body effects on the velocity autocorrelation function but insignificant effect on the self diffusion coefficient (16).

5. Conclusions

A multiple time step method has been successfully developed for performing molecular dynamics simulations in which three body interactions are explicitly included. The method has been tested for 108 particles interacting with a Lennard-Jones plus Axilrod-Teller potential. When compared with a conventional molecular dynamics program using the same potential model, the MTS program executes ~4.5 times faster while giving essentially the same values for equilibrium and dynamic properties, for runs of up to 2000 time steps. At this point, the MTS method is of doubtful reliability for studying long-lived phenomena. Based on this study, it seems that the multiple time step methods could provide a mechanism for extending molecular dynamics simulations to a variety of phenomena which involve components that inherently operate on different time scales; for example, study of phase transitions or long-chain molecules.

In preliminary studies of three body effects with the MTS program, we have confirmed the work of Barker, et al. (11, 12) that the Axilrod-Teller contribution is only a few per cent for the internal energy but significantly larger for the pressure. The effect of the Axilrod-Teller interaction on the radial distirbution function is found to be small. This result is in agreement with Schommers' two dimensional simulations (16); but disagrees with the theoretical calculations of Singh, et al. at high density (14, 15).

Finally, we note that the work reported here begins a study of long range, three body interactions but does not include possible short range, three body effects. P. A. Egelstaff is studying such effects using neutron scattering and Monte Carlo simulation (36).

Acknowledgments

The author is grateful to: W. B. Streett and D. J. Tildesley for valuable discussions on the multiple time step method; H. W. Graben for discussions on three body forces; P. A. Egelstaff and J. A. Barker for instructive correspondence. This work was supported, in part, by a grant from the Faculty Research Committee, Clemson University. The Clemson University Computer Center generously provided the computer time used in this work.

Literature Cited

1. Hansen, J. P. and McDonald, I. R., "Theory of Simple Liquids," Academic Press, New York, 1976.
2. Watts, R. O. and McGee, I. J., "Liquid State Chemical Physics," J. Wiley and Sons, New York, 1976.
3. Metropolis, M., Rosenbluth, A. W., Rosenbluth, M. N., Teller, A. N., and Teller, E., J. Chem. Phys. (1953) 21,1087.
4. Alder, B. J. and Wainwright, T. E., J. Chem. Phys. (1957) 27, 1208.
5. Streett, W. B., Tildesley, D. J., and Saville, G., Molec. Phys. (1978) in press.
6. Copeland, D. A. and Kestner, N. R., J. Chem. Phys. (1968) 49, 5214.
7. Casanova, G., Dulla, R. J., Jonah, D. A., Rowlinson, J. S., and Saville, G., Molec. Phys. (1970) 18, 589.
8. Dulla, R. J., Rowlinson, J. S., and Smith, W. R., Molec. Phys. (1971) 21, 299.
9. Sherwood, A. E. and Prausnitz, J. M., J. Chem. Phys. (1964) 41, 413.
10. Fowler, R. and Graben, H. W., J. Chem. Phys. (1972) 56, 1917.
11. Barker, J. A., Fisher, R. A., and Watts, R. O., Molec. Phys. (1971) 21, 657.

12. Barker, J. A., Henderson, D., and Smith, W. R., Molec. Phys. (1969) 17, 579.
13. Yarnell, J. L., Katz, M. J., Wenzel, R. G., and Koenig, S. H., Phys. Rev. A (1973) 7, 2130.
14. Ram, J. and Singh, Y., J. Chem. Phys. (1977) 66, 924.
15. Sinha, S. K., Ram, J., and Singh, Y., J. Chem. Phys. (1977) 66, 5013.
16. Schommers, W., Phys. Rev. A (1977) 16, 327.
17. Bell, R. J., J. Phys. B (1970) 3, 751.
18. Axilrod, B. M. and Teller, E., J. Chem. Phys.(1943) 11, 299.
19. Muto, Y., Proc. Phys. Math. Soc. Japan (1943) 17, 629.
20. Jansen, L. and Lombardi, E., Faraday Disc. Chem. Soc. (London) (1965) 40, 78.
21. Barker, J. A. and Henderson, D., Rev. Mod. Phys. (1976) 48, 587.
22. Verlet, L., Phys. Rev. (1968) 165, 201.
23. Fisher, R. A. and Watts, R. O., Austr. J. Phys. (1972) 25, 529.
24. Singh, Y., Molec. Phys. (1975) 29, 155.
25. Shukla, K. P., Ram, J., and Singh, Y., Molec. Phys. (1976) 31, 873.
26. Stogryn, D. E., J. Chem. Phys. (1970) 52, 3671.
27. Gear, C. W., "Numerical Initial Value Problems in Ordinary Differential Equations," Prentice-Hall, Englewood Cliffs, New Jersey, 1971.
28. Cheung, P. S. Y. and Powles, J. G., Molec. Phys. (1974) 30, 921.
29. Haile, J. M., "Surface Tension and Computer Simulation of Polyatomic Fluids," Ph.D. Dissertation, University of Florida, Gainesville, 1976.
30. Wood, W. W., in "Physics of Simple Liquids," H. N. V. Temperley, J. S. Rowlinson, and G. S. Rushbrooke, eds, North-Holland, Amsterdam, 1968.
31. Verlet, L., Phys. Rev. (1967) 159, 98.
32. Schofield, P., Comput. Phys. Comm. (1973) 5, 17.
33. Quentrec, B. and Brot, C., J. Comput. Phys. (1973) 13, 430.
34. Rahman, A., Phys. Rev. (1964) 136, 405.
35. Hoover, W. G. and Ashurst, W. T. in "Theoretical Chemistry," vol. 1, H. Eyring and D. Henderson, eds., Academic Press, New York, 1975.
36. Egelstaff, P. A., private communication (1978).

RECEIVED September 7, 1978.

16

Monte Carlo Studies of the Structure of Liquid Water and Dilute Aqueous Solutions

DAVID L. BEVERIDGE, MIHALY MEZEI, S. SWAMINATHAN, and S. W. HARRISON

Chemistry Department, Hunter College of the City University of New York, 695 Park Avenue, New York, NY 10021

The advent of third generation digital computer hardware together with recent advances in computational chemistry based on molecular quantum mechanics and statistical mechanics combine to make the structure of molecular liquids accessible to theoretical study at new levels of rigor via computer simulation. There is presently broad based research activity in this area from quite diverse points of view in physics and chemistry.

A series of Monte Carlo computer simulation studies of the structure and properties of molecular liquids and solutions have recently been carried out in this Laboratory.[1-4] The calculations employ the canonical ensemble Monte Carlo-Metropolis method based on analytical pairwise potential functions representative of ab initio quantum mechanical calculations of the intermolecular interactions. A number of thermodynamic properties including internal energies and radial distribution functions were determined and are reported herein. The results are analyzed for the structure of the statistical state of the systems by means of quasi-component distribution functions for coordination number and binding energy. Significant molecular structures contributing to the statistical state of each system are identified and displayed in stereographic form.

This article reviews the main results of our most recent work and deals specifically with liquid water, the dilute aqueous solution of methane, and dilute aqueous solutions of monatomic cations and anions. The background for these studies is surveyed in Section I, followed by general considerations on the methodology and computational parameters. Sections III-V collect the individual results system by system, followed in Section VI by a general discussion and conclusions.

I. Background

The motional degrees of freedom available to molecules in liquids and solutions mandate theoretical studies of these systems to be problems in statistical mechanics and dynamics. The most fundamental approach to problems in this area is to treat each

0-8412-0463-2/78/47-086-191$07.00/0

system as simply an assembly of molecules interacting under a configurational potential, and to attempt to solve the corresponding many-body problem in classical statistical mechanics or kinetic theory. When the intermolecular interactions are relatively strong, as in most molecular liquids, general solutions cannot be readily developed in analytical form. However, recent studies show it is possible to approach these problems by numerical methods using large scale digital computers. In statistical mechanics, the system can be treated by configurational averaging using the Monte Carlo-Metropolis method, whereas in the kinetic approach, molecular dynamics, the individual molecular trajectories are calculated by simultaneous solution of the Newton-Euler equations. The Monte Carlo and molecular dynamic methods are collectively referred to as "computer simulations". In principle, the computer simulation methods are able to accommodate all the thermodynamic and related properties of an equilibrium system, and molecular dynamics can be extended to non-equilibrium systems as well. The simulation methods also afford the possibility of carrying out computer experiments on the system to elucidate the effects of various definable characteristics on the results. There remain significant assumptions in computer simulation, mainly regarding the proper form of the configurational potential, and individual calculations themselves are relatively lengthy undertakings in terms of computer time. However, the initial results on molecular liquids and solutions are encouraging and are providing new insights into the structural chemistry of these systems.

A large body of the computer simulation work has been reported on model systems such as hard discs, spheres or Lennard-Jones particles. Here the interparticle potential is known and can be used to rapidly calculate the configurational energy of the system as required for Monte Carlo studies or the configurational force on a particle as required for molecular dynamics. A great deal of systematic information has been developed from model systems which can be qualitatively applicable to real systems. A series of definitive reviews of the Monte Carlo method and results on model systems have been prepared by Wood[5]. The dynamics approach was initially characterized in the series of papers by Alder, Wainwright and coworkers[6]. A comprehensive review of liquid state theory was recently published by Barker and Henderson[7].

The application of computer simulation methods to molecular systems is in a relatively early stage. The Monte Carlo studies on water by Barker and Watts (1969)[8] and Sarkisov et. al. (1974)[9] and the molecular dynamics study of water by Rahman and Stillinger (1971)[10] were the forerunners of computer simulation work on chemical problems. Progress has generally been slow in this area due to the magnitude of computer facilities required and the limited availability of potential functions for diverse chemical applications. Quite recently several computer simulation studies on liquid water have appeared, due to Ladd,[11] Owicki and Scheraga,[12] and Swaminathan and Beveridge.[1] Clementi and coworkers have reported extensive results on water[13] and ion-water systems[14] and have pioneered the use of intermolecular potential functions derived from ab initio quantum mechanical calculations. A systematic approach to the determination of analytical potential functions was contributed from this Laboratory.[15] The dynamics of ion-water interactions for small clusters has recently been described, and computer simulations have just been reported on liquid benzene,[16] liquid nitrogen[17] and liquid ammonia.[18] The dilute aqueous solution of methane has been treated by Owicki and Scheraga[19] and Swaminathan, Harrison and Beveridge.[3] Additional current applications to molecular liquids are in the newly published monograph by Watts[20] and of course in the companion papers in this volume.

Applications of computer simulation methods to biochemical and biological problems are just now beginning to appear. Several groups are interested in studies of polypeptide and protein conformation based on computer simulation. Clementi and coworkers are working on water-amino acid potential functions and related problems dealing with molecular assemblies.[21] Scheraga and co-workers are involved in complementary studies of water[12] and methane-water solution,[19] and are exploring the use of Monte Carlo calculations in other problems of biophysical interest. Karplus is currently using molecular dynamics to study protein folding in solution.[22] The opportunities for theoretical studies of chemical and biological processes using computer simulation are extremely broad and diverse, and we fully expect that this approach will ultimately have a broad impact on theoretical chemistry, biochemistry and biology.

Each of the systems individually under consideration in this review has an extensive scientific history. The background appropriate to each system is developed in considerable detail in our individual papers and, except for particular points and references to the most recent relevant work, is not repeated here. The focus herein is thus on presentation of collected current Monte Carlo results from this Laboratory obtained on a series of important systems, with the computational procedures and analyses carried out on a unified and coherent basis.

II. Calculations

The methodology of Monte Carlo calculations especially for polar and ionic systems is currently an active area of study in computer simulation theory. Aspects such as potential functions, boundary conditions, sampling algorithms, convergence criteria and truncation errors are all receiving considerable research attention. The sensitivity of results to assumptions in the calculations is discussed particularly in a recent paper by Levesque, Patey and Weiss.[23]

All calculations described herein are based on statistical thermodynamic computer simulation under canonical ensemble conditions on the system, with temperature T, number of particles N, and volume V assumed and constant. Configurational integrations for each system are carried out by Monte-Carlo methods involving a stochastic walk through configuration space with configurations selected on the basis of their probability in the ensemble by the method suggested by Metropolis et. al.[24] The N-particle system is given a liquid phase environment by the appropriate choice of number density and the use of image cells, i.e. periodic boundary conditions. The formalism employed is developed fully in References 1 and 3. All calculations unless otherwise indicated are based on 125 particles in a cubic cell at a temperature at 25°C. The volume of the cell is determined by density and specified individually for each system. The calculations involve a spherical cutoff in the potential function at half the cell edge used in conjunction with the minimum image convention. Convergence criteria and error bounds on each of the calculated quantities in the simulation are developed in terms of control functions as defined by Wood.[5]

The configurational energy of the system in each of the Monte Carlo calculations discussed below is developed under the assumption of the pairwise additivity of intermolecular interactions by means of analytical functions representative of ab initio quantum mechanical calculations of the intermolecular interaction energy. There are of course a number of limiting assumptions involved in the construction of such functions, such as the size and specification of basis sets in the quantum mechanical calculations, but they are formally well defined and their capabilities and limitations with respect to the pairwise interaction can be developed in detail. The functions can be readily defined on the requisite regions of configuration space, and anisotropies in the intermolecular interactions are of course automatically included. The individual terms in the analytical potential function should not be ascribed any physical significance; they are simply a means for an interpolation based on a limited number of discrete quantum-mechanically calculated intermolecular interaction energies. The major assumptions inherent in the configurational energy calculations in this study are thus the neglect of three-body and higher order contributions, truncation errors in the quantum mechanical calculations of pairwise interaction energies, and statistical errors in the multidimensional curve fitting in the analytical potential.

All calculated quantities reported, with the exception of the free energy for liquid water, are based on simple ensemble averages and are produced in a straightforward manner in the Metropolis procedure. The analysis of results is based particularly on quasicomponent distribution functions as introduced by Ben-Naim.[25] Quasicomponent distribution functions are defined on the statistical state of the system and give the distribution of particles (in terms of mole fraction) with any particular well defined characteristic. The characteristics relevant to this work are a) coordination number, the number of particles within a given radius R_M of the center of mass of a given particle, and b) binding energy, the interaction energy of a given particle with all other particles in the system. The quasicomponent distributions are of course obtained as averages over all identical particles and all configurations of the system.

III. Liquid Water

The extensive use of water as a solvent in chemical systems, the direct participation of water in many chemical and biochemical processes and the unique function of water as a biological life support system combine to make the structure of liquid water a matter of central importance in understanding many chemical, biochemical and biological systems at the molecular level. Liquid water has been studied extensively using statistical thermodynamics based on ad hoc models for the system, and there has been a long-standing controversy in the scientific literature as to whether the structure of water is best represented by a mixture model, whereby the system is viewed as a composite of energetically distinct clusters of possibly diverse size and structure, or by an energetic continuum model, wherein local molecular environments with a continuous distribution of progressively bent hydrogen bonds are featured. Arguments pro and con for each model based on both theoretical analysis and experimental evidence are summarized in the recent reviews by Davis and Jarzynski[26] and Kell[27] and general perspective on the problem is developed in the series edited by Franks[28] and a recent review by Gorbunov and Naberukhin.[29]

Liquid water was the first molecular liquid to be extensively studied using computer simulation methods. Experimental data on diverse thermodynamic properties are available and Narten, Levy and co-workers have obtained radial distribution functions for the system by diffraction methods.[30] This system thus serves as a sensitive test on the quality of the configurational potential used in a computer simulation. The calculated oxygen-oxygen radial distribution function of liquid water has been reported for quite a variety of pairwise additive empirical and analytical potential functions. Among the empirical functions, the ST2 potential[31] is currently widely adopted. Clementi and co-workers[13] carried out extensive studies of pairwise potential functions representative of *ab initio* quantum mechanical calculations on the water dimer, and best agreement with the observed radial distribution function was obtained using a function representative of a self consistent field (SCF) calculation plus moderately large intermolecular configuration interaction (CI) on the system. The function was reported by Matsuoka, Clementi and Yoshimine (MCY)[32] and is henceforth referred to herein as the MCY-CI potential.

Recent studies on liquid water in this Laboratory[1] have been directed toward an analysis of the structure of the system from Monte Carlo computer based on the MCY-CI potential function. The calculations were carried out for 125 particles at a temperature of 25°C and a volume commensurate with a density of 1 gm/cm^3. The ensemble averages for the system were based on 500K configurations selected on the basis of the Metropolis method. Convergence was checked by comparing the results of two independent runs, one beginning at a configuration with relatively high energy and another beginning at a configuration very near the minimum energy for the system. The calculations converged to the same internal energy within 1% and apparently no practical nonergodicities were encountered.[33]

The calculated thermodynamic internal energy for the system obtained from Monte Carlo computer simulation based on the MCY-CI potential was -8.58 ± 0.06 kcal/mol, compared with an observed value of -9.9 kcal/mol.[9] The discrepancy of ~13% is ascribed mainly to the assumption of pairwise additivity in the configurational potential. The calculated heat capacity, corrected for internal modes, is 17.9 cal/mol · deg as compared with the experimental value at 25° of 18 cal/mol · deg.

The calculated oxygen-oxygen radial distribution function for liquid water obtained from Monte Carlo computer simulation based on the MCY-CI potential is shown in Fig. 1. There is quite good accord between the calculated and observed values for the position of all three main peaks. In the region of the first hydration shell, the shape of the calculated peak agrees well with experiment, although the calculated maximum is slightly too high. For the second shell, the position of the maximum is correct but the shape is biased towards short distance side. A shoulder at ca. 3.5 Å is clearly indicated. The third shell appears generally well described.

An analysis of the structure of liquid water was carried out in terms of quasicomponent distribution functions for coordination number and binding energy and also by examining stereographic views of significant molecular structures. Coordination number is calculated based on $R_M = 3.3$, the first minimum in the radial distribution function, and this distribution function is essentially an analysis of the first hydration shell. The calculated result, displayed as mole fraction of particles $x_C(K)$ vs. coordination number K is shown in Figure 2. The distribution ranges from K=2 to K=6, biased towards higher coordination numbers, with K=4 predominant at 47%. The average coordination number was found to be 4.1.

The quasicomponent distribution function for binding energy, displayed as mole fraction of particles $x_B(\nu)$ vs. binding energy ν, is given in Figure 3. The distribution is unimodal with a maximum at -17.7 kcal/mole. As discussed by Ben-Naim,[25] a mixture model for the system should give rise to a bimodal or poly-

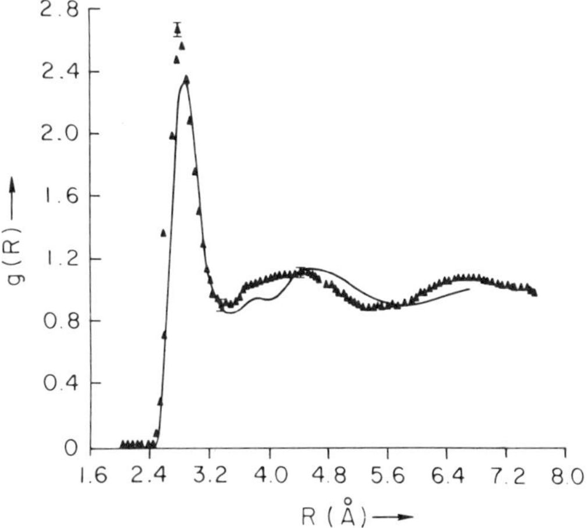

Figure 1. Calculated points on the oxygen–oxygen radial distribution function g(R) *vs. center of mass separation* R *for liquid water at 25°C. Experimental data (solid line) from Narten, Danford, and Levy* (30).

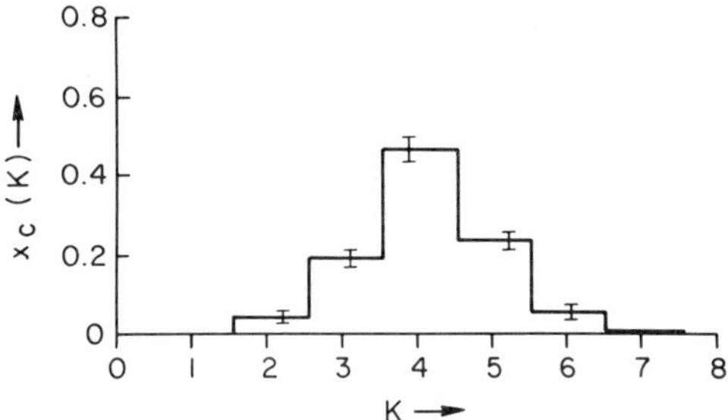

Figure 2. Calculated quasicomponent distribution function $x_C(K)$ *vs. coordination number* K *for liquid water*

modal distribution of binding energies while a unimodal distribution is consistent with the idea of an energetic continuum model. Thus the modality of the calculated distribution function for binding energy supports the continuum model for liquid water. Examining these results from the point of view of hydrogen bond energy, an ice-like environment with four linear hydrogen bonds would be expected to have a binding energy of ~22 kcal/mol. This energy is on the lower edge of the calculated distribution and not heavily favored. The region of higher probability must therefore correspond to bent hydrogen bonds. To examine this point further, we extracted from the calculations a number of low energy, high frequency structures. Stereographic views of two of them, chosen such that the central molecule has K=4, are shown in Figures 4 and 5. There is a notable prevalence of bent hydrogen bonds throughout the structure.

Collectively, these results support the energetic continuum model for liquid water. Similar conclusions emerged from the previously mentioned molecular dynamics study of the system. In other theoretical work on this problem, Kauzmann[34] has demonstrated that a 2-state mixture model for the system cannot formally account for all the observed data on the system. Vibrational spectral data supporting a mixture model based on an isosbestic point in the Raman spectra as a function of temperature have been reported by Walrafen.[35] This problem has been reexamined recently by Scherer, Go and Kint[36] who showed that the apparent isosbestic point in Raman data that has not been decomposed into isotropic and anisotropic parts is fortuitous. Rice[37] and coworkers find their detailed analysis of the O-H stretching region of the vibrational spectra of amorphous solid water consistent with a slightly bent hydrogen bond model. The relationship between the observed far infared spectrum of water and the molecular dynamics computer simulation results has been developed by Curnette and Williams.[38] The recent reconsideration of spectroscopic aspects of the water structure problem due to Gorbunov and Naberukhin[29] firmly supports the continuum model.

The Monte Carlo-Metropolis method for computer simulation is ideally suited to produce ensemble averages of the properties of the system. However in the case of free energy, the ensemble average expression is not convenient for computational purposes due to the ill-conditioned nature of the integrand and the concomitant convergence problems.[19] We have recently carried out a Monte Carlo calculation of the free energy of liquid water based on a procedure which follows from early work by Kirkwood[39] involving a numerical quadrature over the internal energy of the system developed in terms of an auxiliary parameter.[2] The calculation was found to be computationally tractable and resulted in a Helmholz configurational free energy of -4.31 ± 0.07 kcal/mol compared with an observed value of -5.74 kcal/mol;[9] the corresponding entropy was found to be -14.44 ± 0.09 cal/deg mol against an observed value of -13.96 cal/deg mol at 25°C. The error in free energy being

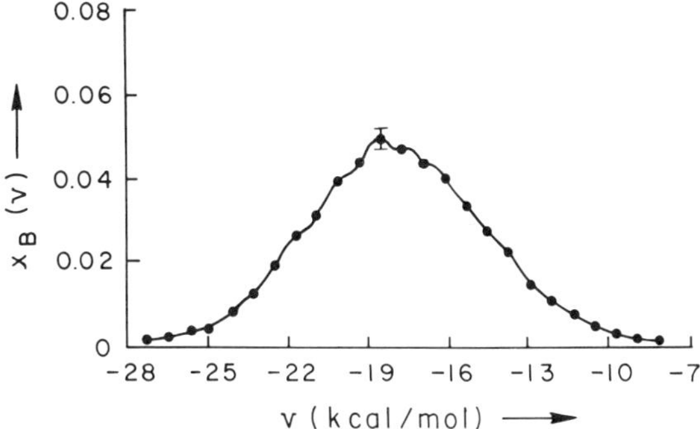

Figure 3. Calculated quasicomponent distribution function $x_B(v)$ vs. binding energy v for liquid water

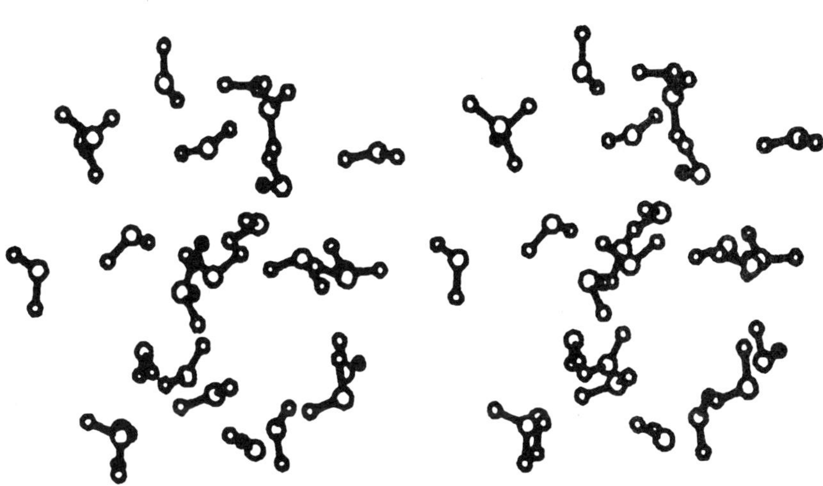

Figure 4. Stereographic view of a fragment of a significant molecular structure contributing to the statistical state of liquid water

of the same order of magnitude as the error in the internal energy, we ascribe this discrepancy also to the neglect higher order contributions to the configurational potential.

IV. The Dilute Aqueous Solution of Methane

The dilute aqueous solution of methane is a system of prominent interest in molecular liquids as the prototype of a non-polar molecular solute dissolved in liquid water, and is one of the simplest molecular systems where the hydrophobic effect[40] is manifest. Moreover, a detailed knowledge of the structure of the methane-water solution at the molecular level can provide leading information on the interaction of water with dissolved hydrocarbon chains in general, and thereby contribute to the theoretical basis for understanding the role of water in maintaining the structural integrity of biological macromolecules in solution.

General background on solutions of nonpolar solutes in water has been recently reviewed by Franks[41] and Ben-Naim.[25,42] Early important work on these systems is due to Eley[43] and Frank and Evans.[44] Methane has been identified as a "structure-maker" in aqueous solution in the language of Frank and Wen.[45] The nature of structural changes in solvent water by dissolved hydrocarbons has for sometime been discussed as by Tanford[40] in terms of water clathrate formation, based on work particularly by Glew[46] and analogies drawn from a number of hydrate crystal structures of non-polar species, known to involve water clathrate structures of order 20 and 24. Key papers on this topic include the review by Kauzmann[47] and work by Scheraga and coworkers[48].

Early computer simulation work on the methane-water system was reported by Dashevsky and Sarkisov.[49] Recent theoretical studies of the methane-water system are the *ab initio* molecular orbital calculations of the methane-water pairwise interaction energy by Ungemach and Schaefer[50] and the Monte Carlo computer simulation on the dilute aqueous solution in the isothermal-isobaric ensemble by Owicki and Scheraga.[19]

A recent Monte Carlo study of structure of the dilute aqueous solution of methane from this Laboratory[3] involves one methane molecule and 124 water molecules at 25°C at liquid water density. The configurational energy of the system is developed under the assumption of pairwise additivity using potential functions representative of *ab initio* quantum mechanical calculations for both the water-water and methane-water interactions. For the water-water interaction we have carried over the MCY-CI potential function used in our previous study of the structure of liquid water reviewed in the preceeding section. For the methane water interaction energy, we have recently reported[51] an analytical potential function representative of quantum mechanical calculations based on SCF calculations and a 6-31G basis set, with correlation effects included via second order Moller-Plesset (MP) corrections,[52] This function was used for the methane-water contributions to the configurational potential presented herein. Ensemble averages were calculated on the basis of a 650K stochastic walk.

The calculated partial molar internal energy for transfer of methane from gas phase to water was -23.3 ± 6.6 kcal/mol, compared with an experimental value of -2.6 kcal/mol.[53] The calculated result is seen to be negative as expected for the hydrophobic effect, but an order of magnitude too low. This discrepancy is discussed further below. The calculated radial distribution function for the center of mass of water molecules with respect to the methane carbon atom is shown in Figure 6. We find a broad, unstructured first peak with a minimum in the region of 5.3Å. Integrating g(R) up to this point gives an average coordination number of 19.35.

An analysis of the structure of the dilute aqueous solution of methane was also developed in terms of quasicomponent distribution functions and stereographic views of significant molecular structures. The coordination number of methane in this system was calculated on the basis of $R_M = 5.3$Å, fixed at the first minimum in the methane-water radial distribution function. A plot of the mole fraction of methane molecules $x_C(K)$ vs. their corresponding water coordination number is given in Figure 7. The $x_C(K)$ obtained is a broad unimodal distribution ranging from K=16 to K=22 with a maximum in the region K=19 and 20, biased slightly in shape towards higher coordination numbers.

The calculated quasicomponent distribution function for binding energy, the mole fraction of methane molecules $x_B(\nu)$ as a function of methane binding energy ν is shown in Figure 8. There is some incipient structure in the curve but the error bounds on the function are too large to ascribe this any physical significance. Comparing the average value of this quantity, -23.3 ± 6.6, with the calculated partial molar internal energy of transfer for methane confirms that the sign and magnitude of this latter term are due to water stabilization effects. If we now assume the $\sim$20 kcal/mole discrepancy between the calculated and observed values of the internal energy comes from the $\sim$20 water molecules found in the first hydration shell of methane, this is an error of $\sim$1 kcal/mole per water molecule or a fraction of a kcal/mol per pairwise interaction or hydrogen bond. Considering the approximations inherent in the calculation of pairwise interaction energies as enumerated in Section II, this discrepancy appears to be commensurate with the capabilities and limitations of the configurational energy calculation.

The magnitudes and distribution of coordination numbers found for methane in the statistical state of the dilute aqueous solution are generally consistent with the presence of water clathrate cages as shown in Figure 9. In a number of other structures quasiclathrate regions could be identified, but defects and distortions were more prevalent than in Figure 9. One such structure, representative of other, is given in Figure 10. The

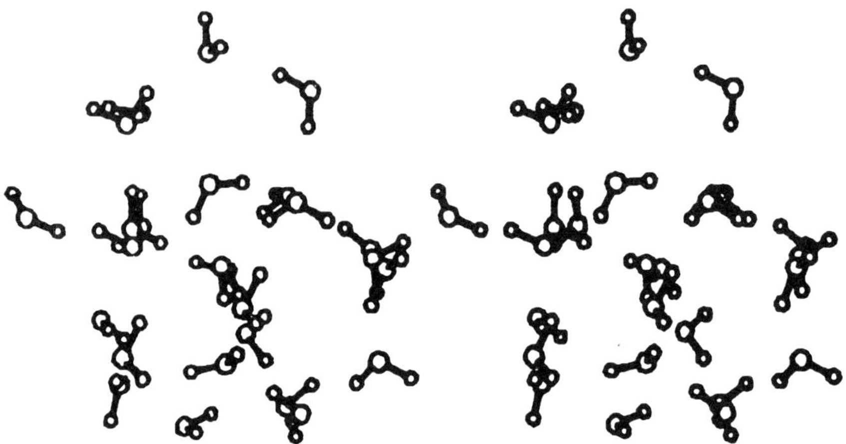

Figure 5. Stereographic view of a fragment of another significant molecular structure contributing to the statistical state of liquid water

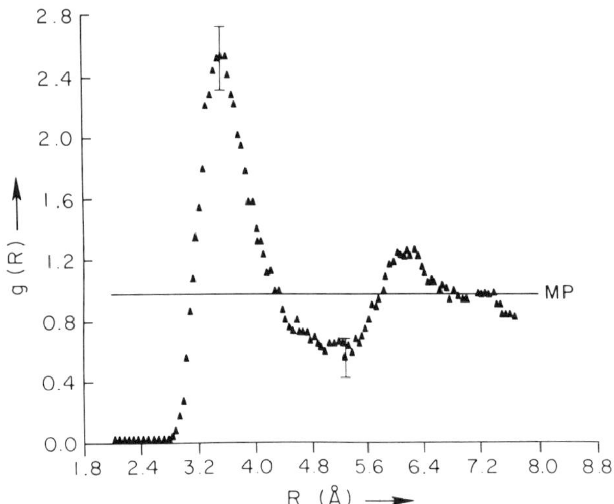

Figure 6. Calculated methane–water radial distribution g(R) *vs. center of mass separation* R *from Monte Carlo computer simulation for the dilute aqueous solution of methane at* T = 25°C

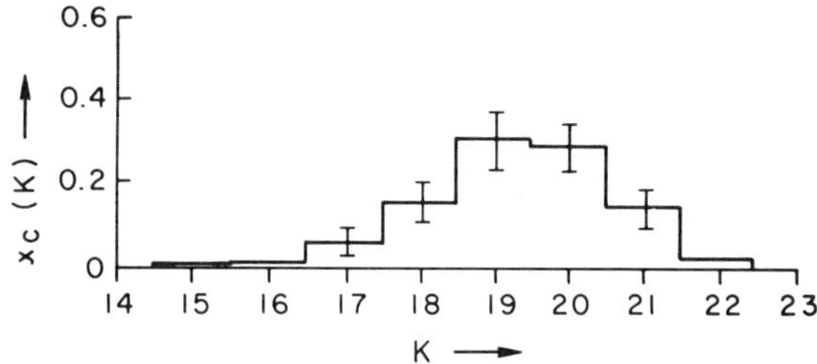

Figure 7. Calculated quasicomponent distribution function $x_C(K)$ *vs. methane coordination number* K *for the dilute aqueous solution of methane*

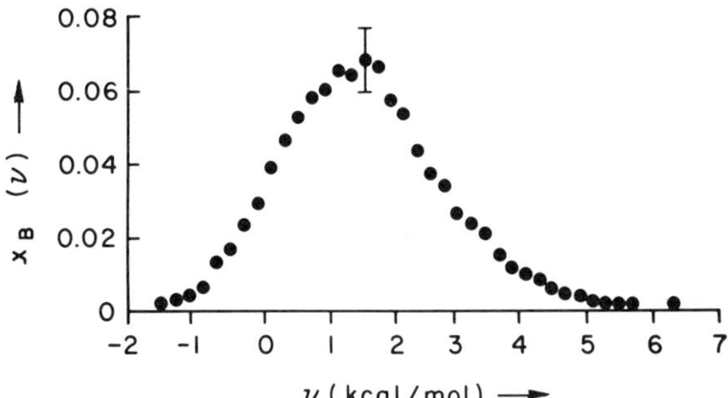

Figure 8. Calculated quasicomponent distribution function $x_B(\nu)$ *vs. methane binding energy* ν *for the dilute aqueous solution of methane*

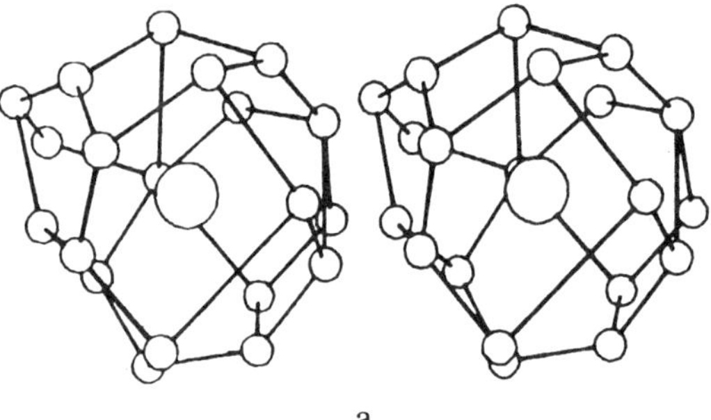

a

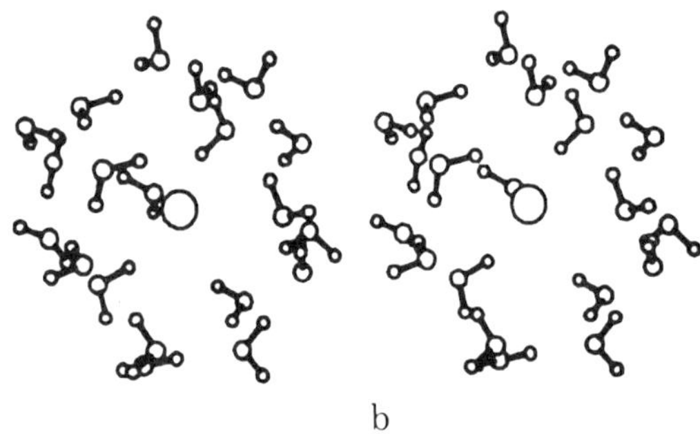

b

Figure 9. Stereographic view of methane and its first hydration shell taken from a significant molecular structure contributing to the statistical state of the system. (Top) *disposition of centers of mass of water molecules about methane (shaded) with the quasiclathrate cage delineated;* (bottom) *disposition of water molecules about methane in the same structure.*

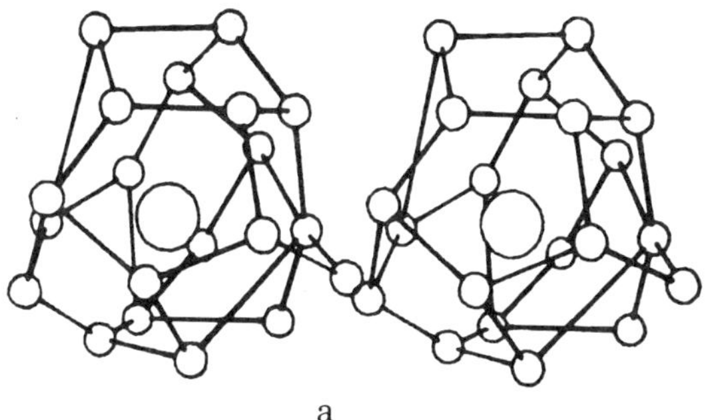

a

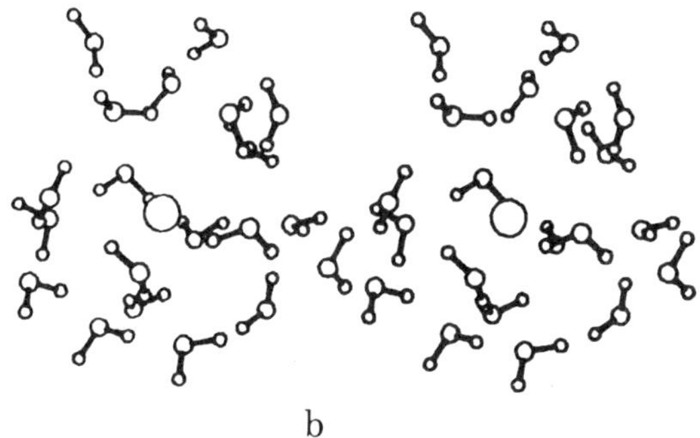

b

Figure 10. Stereographic view of methane and its first hydration shell taken from another significant molecular structure contributing to the statistical state of the system. (Top) *disposition of centers of mass of water molecules about methane (shaded) with the quasiclathrate cage delineated;* (bottom) *disposition of water molecules about methane in the same structure.*

emergent description of the aqueous solution environment of methane is that of a distorted and defective continuum clathrate structure.

Effects of solute methane on solvent water structure were developed in terms of the difference in $x_C(K)$ and $x_B(\nu)$ calculated for the solution and for the pure liquid. The effects of bulk water are removed by the differencing, allowing the direct display of structural changes in solvent water. The difference plots reveal under high statistical error an increase in 4-coordinate species and a slight but clearly discernable shift toward lower binding energy for the solvent molecules, provisionally consistent with general ideas of "structure making". Further work is currently in progress on this point.

V. Dilute Aqueous Solutions of Monatomic Cations and Anions.

The particular significance of electrolytes in solution chemistry makes the structure of dilute aqueous solutions of monatomic cations and anions also a topic of fundamental interest. Moreover, the sodium and potassium ions in particular figure prominently in biochemical membrane potential phenomena, and their hydration state in aqueous solution is an important factor in ion selectivity and membrane permeability in biological systems.

Modern theoretical studies of these systems date from the classic paper of Bernal and Fowler in 1933.[54] The current state of both experimental and theoretical research on ionic solutions is the subject of several recent comprehensive reviews particularly by Friedman and co-workers.[55] The prevalent descriptive ideas about the local aqueous solution environment of ions stems from the work of Frank and Wen,[45] who partitioned the solvent into three regions. In the immediate vicinity of the ion, region A, water molecules are tightly bound and highly oriented. Region C at large distance from the ion was considered as essentially bulk water, and the intervening region B is a region of structural ambiguity interfacing regions A and B.

Recently Kistenmacher, Popkie and Clementi[56] reported analytical intermolecular potential functions for ion-water interactions representative of near Hartree-Fock molecular orbital calculations. These functions along with water-water potentials of commensurate quality were used to study the structure and solvation numbers of ion-water clusters based on energy optimization. Monte Carlo simulation work based on these functions have been reported by Mruzik, Abraham, Scheiber and Pound[57] focussing on free energy calculations, and by Watts[20] and coworkers on ion pairs in large water clusters. McDonald and Rasaiah have carried out very large order studies of ion pairs in solution based on model potential functions. Molecular dynamics on ion-water systems have been reported by Briant and Burton[59] and Heinzinger and Vogel.[60]

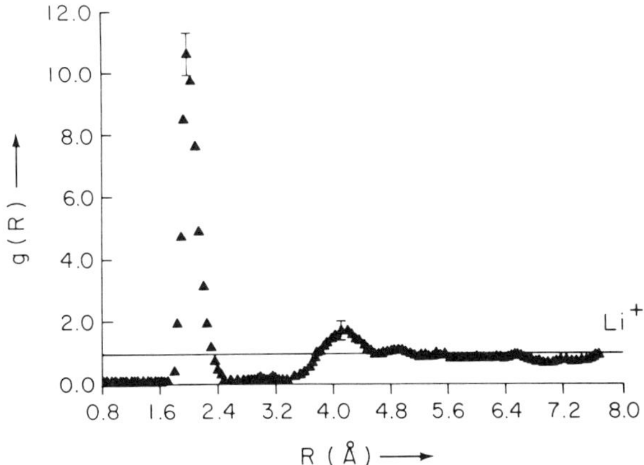

Figure 11. Calculated cation–water radial distribution function vs. center of mass separation R *for the dilute aqueous solution of lithium at* T = 25°C

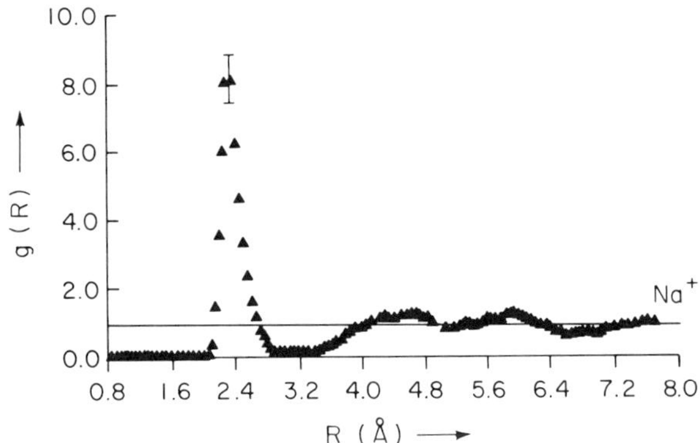

Figure 12. Calculated cation–water radial distribution function vs. center of mass separation R *for the dilute aqueous solution of sodium at* T = 25°C

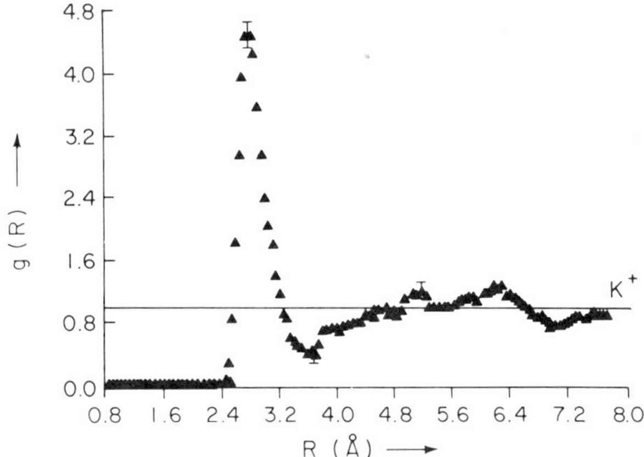

Figure 13. Calculated cation–water radial distribution function vs. center of mass separation R *for the dilute aqueous solution of potassium at* T = 25°C

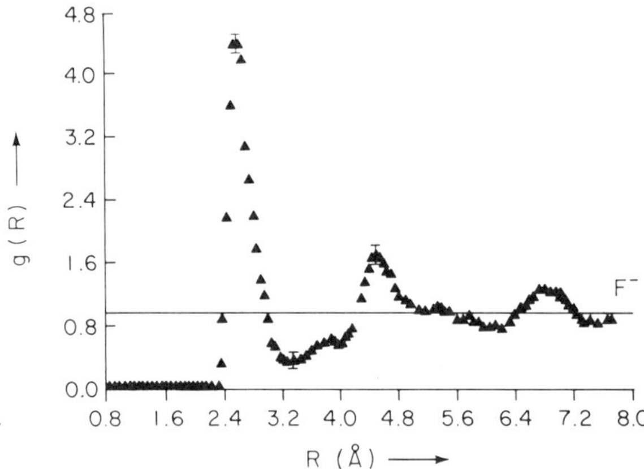

Figure 14. Calculated anion–water radial distribution function vs. center of mass separation R *for the dilute aqueous solution of fluoride at T = 25°C*

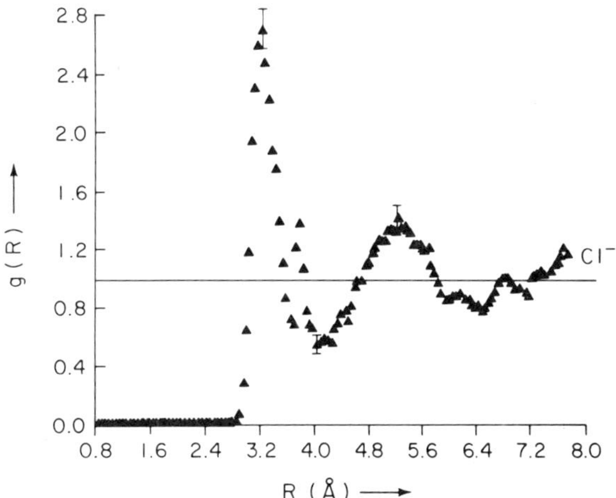

Figure 15. Calculated anion–water radial distribution function vs. center of mass separation R *for the dilute aqueous solution of chloride at* T = 25°C

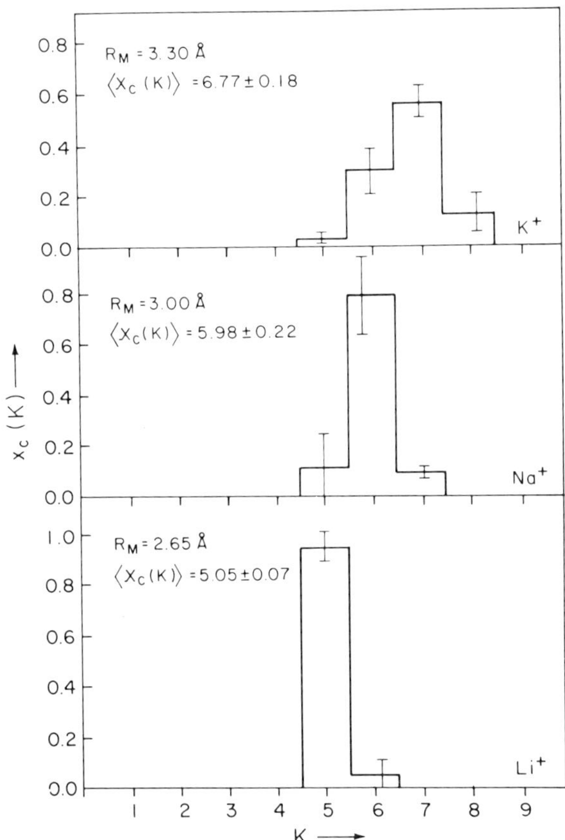

Figure 16. Calculated quasicomponent distribution functions $x_C(K)$ *vs. ion coordination number* K *for dilute aqueous solutions of alkali metal cations*

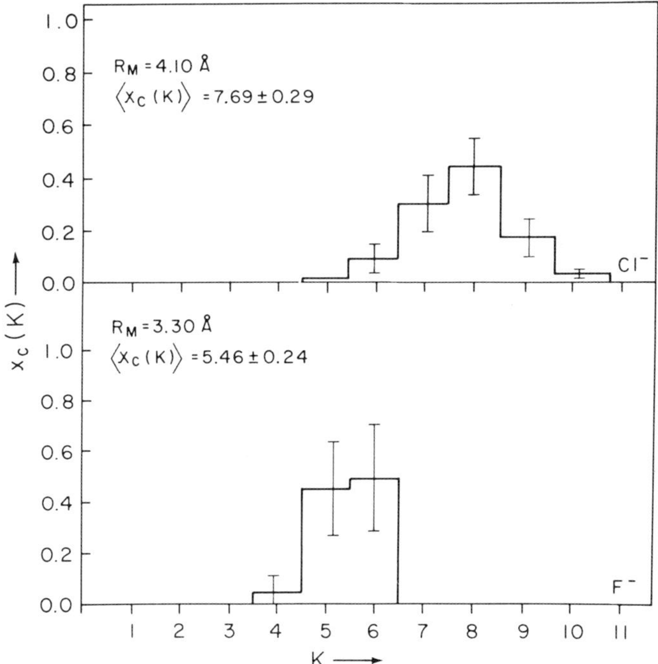

Figure 17. Calculated quasicomponent distribution functions $x_C(K)$ *vs. ion coordination number* K *for dilute aqueous solutions of halide anions*

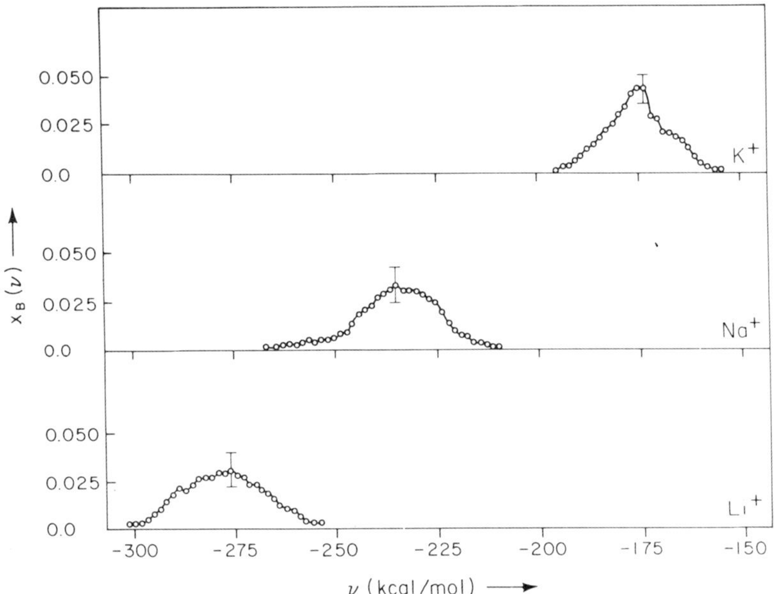

Figure 18. Calculated quasicomponent distribution functions $x_B(\nu)$ vs. ion binding energy ν for dilute aqueous solutions for alkali metal cations

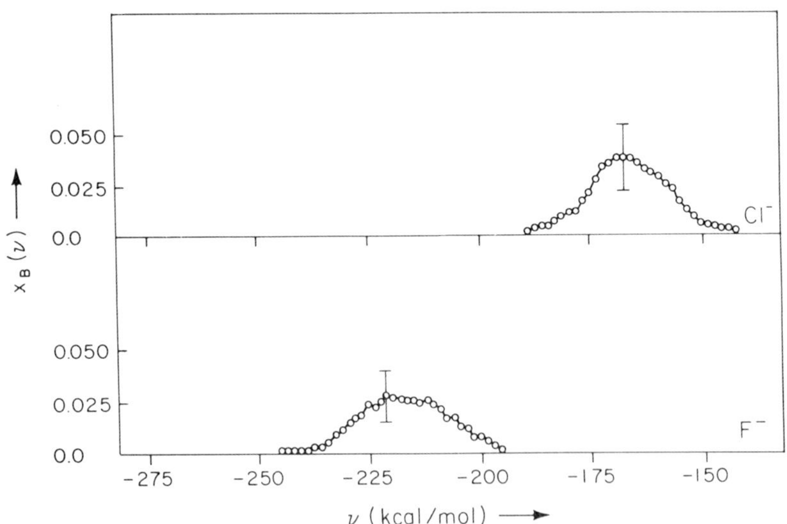

Figure 19. Calculated quasicomponent distribution functions $x_B(\nu)$ vs. ion binding energy ν for dilute aqueous solutions for halide anions

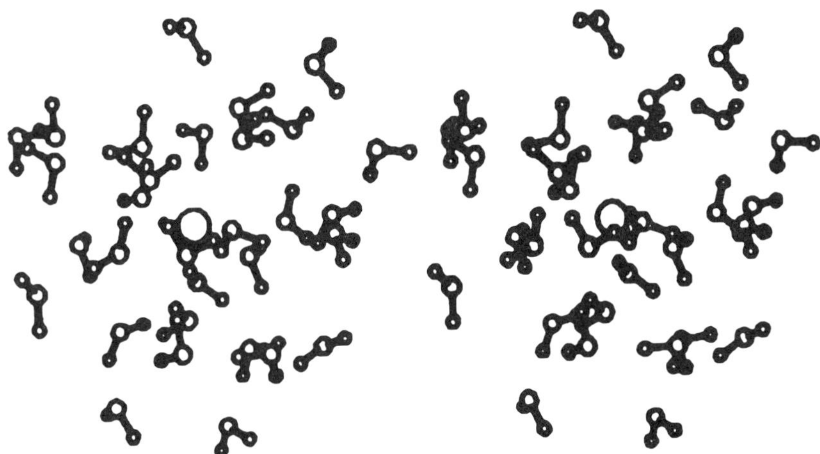

Figure 20. Stereographic view of a significant molecular structure contributing to the statistical state of the dilute aqueous solution of K^+

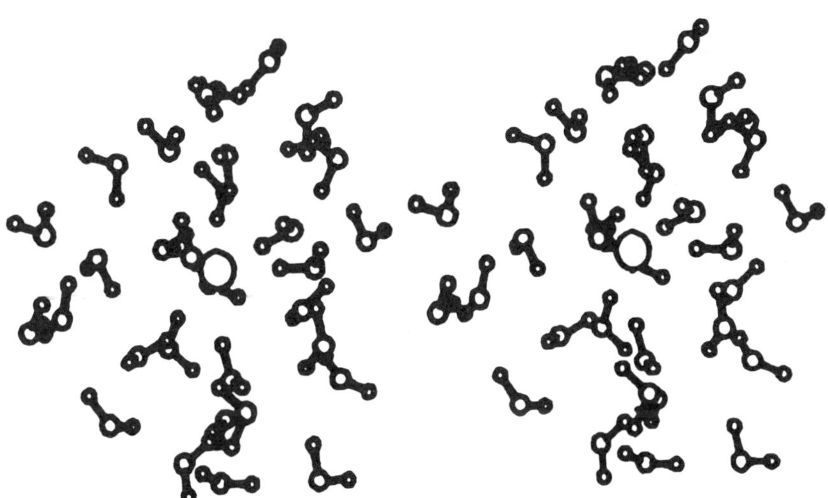

Figure 21. Stereographic view of a significant molecular structure contributing to the statistical state of the dilute aqueous solution of F^-

We have recently carried out Monte Carlo computer simulation of dilute aqueous solutions of the monatomic cations Li^+, Na^+ and K^+ and the monatomic anions F^- and Cl^- using the KPC-HF functions for the ion-water interaction and the MCY-CI potential for the water-water interaction. The temperature of the systems was taken to be 25° and the density chosen to be commensurate with the partial molar volumes as reported by Millero.[61] The calculated average quantities are based on from 600- 900K configurations after equilibration of the systems. The calculated ion-water radial distribution functions are given for the dilute aqueous solutions of Li^+, K^+, Na^+, F^- and Cl^- in Figures 11-15, respectively.

An analysis of the structure of the dilute aqueous solutions of monatomic cations and anions was developed in a manner analogous to that presented above for pure water and the dilute aqueous solution of methane. The distribution of coordination numbers $x_C(K)$ vs. K is presented for the cations in Figure 16 and for the anions in Figure 17. The distributions are all narrow and centered in the region K=5, 6 and 7. The average water coordination number of each ion is found by determining the integral of the first peak in the ion-water radial distribution function, and these values are also recorded on Figures 16 and 17. The binding energy analyses are given in Figures 18 and 19 for cations and anions respectively; each one is continuous and unimodal. Representative stereographic structures are given for the cation K^+ in Figure 20 and for the anion F^- in Figure 21; the other cations and anions introduce no essentially new qualitative features of the structure.

These results should be considered in the context of observed discrepancy of 10-40% between the calculated and observed internal energy of transfer due to the assumption of pairwise additivity in the configurational potential and the truncation of the potential in configurational energy calculations. The effect of higher order terms in the configurational potential on the analysis of structure as presented herein has yet to be determined.

VI. Summary and Conclusions

The series of studies of molecular liquids presented herein collect results on a diverse set of chemically relevant systems from a uniform theoretical point of view: ab initio classical statistical mechanics on the (T,V,N) ensemble with potential functions representative of ab initio quantum mechanical calculations of pairwise interactions and structural analysis carried out in terms of quasicomponent distribution functions. The level of agreement between calculated and observed quantities is quoted to indicate the capabilities and limitations to be expected of these calculations and in that perspective we find a number of structural features of the systems previously discussed on an empirical basis to be quantified. The present results point particularly to a) a continuum model for liquid water, b) a distorted, defective clathrate environment for methane in aqueous solution, and c) distinctly different distributions of equilibrium coordination numbers for Na^+ and K^+ in water which may be of significance in biomembrane phenomena. Ultimately it is hoped that a descriptive chemistry of molecular liquids useful in diverse applications can be developed on a rigorous basis using the results of computer simulations and we contribute these studies as a step in that direction.

VII. Acknowledgement

Support for this research comes from NIH Grant #1-R01-NS12149-03 from the National Institutes of Neurological and Communicative Diseases and Stroke and a CUNY Faculty Research Award. D.L.B. acknowledges U.S.P.H.S. Research Career Development Award 6TK04-GM21281 from the National Institute of General Medical Studies.

VIII. References

1. S. Swaminathan, and D.L. Beveridge, J. Am. Chem. Soc., 99, 8392 (1977).
2. M. Mezei, S. Swaminathan, and D.L. Beveridge, J. Am. Chem. Soc., 100, 3255 (1978).
3. S. Swaminathan, S.W. Harrison, and D.L. Beveridge, J. Am. Chem. Soc., in press.
4. M. Mezei and D.L. Beveridge, MS in preparation.
5. W.W. Wood and J.J. Erpenbeck, Am. Rev. Phys. Chem., 27, 319 (1976). J.J. Erpenbeck and W.W. Wood, in "Statistical Mechanics, Pt B", B.J. Berne, ed., Plenum Press, New York, N.Y. 1977 p 1.
6. B.J. Alder and T.E. Wainwright, J. Chem. Phys., 31, 459 (1959); 33, 1439 (1960). B.J. Alder, D.M. Gass, and T.E. Wainwright, J. Chem. Phys., 53, 3013 (1970).
7. J.A. Barker and D. Henderson, Rev. Mod. Phys., 48, 587 (1976).
8. J.A. Barker and R.O. Watts, Chem. Phys. Lett., 3, 144 (1969).
9. G.N. Sarkisov, V.G. Dashevsky, and G.G. Malenkov, Mol. Phys. 27, 1249 (1974).
10. A. Rahman and F.H. Stillinger, J. Chem. Phys., 55, 336 (1971).
11. A.J.C. Ladd, Mol. Phys., 33, 1039 (1977).
12. J.C. Owicki and H.A. Scheraga, J. Am. Chem. Soc., 99, 7403 (1977).
13. E. Clementi, "Liquid Water Structure", Springer Verlag, New York (1976).

14. J. Fromm, E. Clementi, anf R.O. Watts, J. Chem. Phys., 62, 1388 (1975). E. Clementi, R. Barsotti, J. Fromm and R.O. Watts, Theoret. Chem. Acta (Berl.) 43, 101 (1976).
15. S. Swaminathan, R.J. Whitehead, E. Guth and D.L. Beveridge, J. Am. Chem. Soc., 99, 7817 (1977).
16. D.J. Evans and R.O. Watts, Mol. Phys., 32, 93 (1976).
17. P.S.Y. Cheung and J.G. Powles, Mol. Phys., 32, 1383 (1976).
18. I.R. McDonald and M.L. Klein, J. Chem. Phys., 64, 1790 (1976); P.K. Mehrotra and D.L. Beveridge, MS. in preparation.
19. J.C. Owicki and H.A. Scheraga, J. Am. Chem. Soc., 99, 7413 (1977).
20. R.O. Watts, "Liquid State Chemical Physics", John Wiley & Sons, New York (1976).
21. E. Clementi, F. Lavallone, and R. Scordamoglia J. Am. Chem. Soc., 99, 5531 (1977); R. Scordamoglia, F. Cavallone, and E. Clementi, J. Am. Chem. Soc., 99, 5545 (1977); G. Bolis and E. Clementi, J. Am. Chem. Soc., 99, 5550 (1977).
22. P. Rossky, M. Karplus, private communication, MS in preparation.
23. D. Levesque, G.N. Patey and J.J. Weis, Mol. Phys. 34, 1077 - 1091 (1977).
24. N.A. Metropolis, A.W. Rosenbluth, M.N. Rosenbluth, A.H. Teller and E. Teller, J. Chem. Phys., 21, 1087 (1953).
25. A. Ben-Naim, "Water and Aqueous Solutions", Plenum Press, New York, N.Y. 1974.
26. C.M. Davis and J. Jarzynski in "Water and Aqueous Solutions" R.A. Horne, ed., John Wiley and Sons, New York, N.Y. 1972. The mixture model emerged from the earliest thoughts about the structure of liquid water as a composite of ice-like (bulky) contributions in equilibrium with a dense component. Highly developed forms of this theory are the interstitial model, (O.Ya. Samoilov, Zh. Fiz. Khim. 20, 1411 (1946). V.A. Mikhailov, Zh. Struct. Khim. 8, 189 (1967), 9,397 (1968); A.H. Narten and H.A. Levy, Science 165, 3892 (1969). and the flickering cluster model, (A.T. Hagler, H.A. Scherga and G. Nemethy, Ann. N.Y. Acad. Sci. 204, 51 (1973).
27. G.S. Kell in "Water and Aqueous Solutions" R.A. Horne, ed., John Wiley and Sons, New York, N.Y. 1972. The energetic continuum model for liquid water was developed during dissertation research by J.A. Pople, Proc. Roy. Soc. (London), Ser A, 205, 163 (1951).
28. F. Franks, ed., "Water-A Comprehensive Treatise" Vols. I-III, Plenum Press, New York, N.Y. 1972-73.
29. B.Z. Gorbunov and Yu.I. Naberukhin, J. Struct. Chem. 16, 703 (1975).
30. A.H. Narten and H.A. Levy, J. Chem. Phys., 55, 2263 (1971). A.H. Narten, J. Chem. Phys., 55, 5681 (1972).
31. F.H. Stillinger and A. Rahman, J. Chem. Phys., 60, 1545 (1974).

32. O. Matsuoka, E. Clementi, and M. Yoshimine, J. Chem. Phys., 64, 1351 (1976).
33. B. Larsen and S.A. Rodge, J. Chem. Phys., 68, 1309 (1978). C. Pangali, M. Rao, and B.J. Berne, Chem. Phys. Lett., in press.
34. W. Kauzmann, Colloqes International du C.N.R.S. #2461, 63 (1976).
35. G. Walrafen in "Water - A Comprehensive Treatise", Vol. I, F. Franks, ed., Plenum Press, New York. 1972, p.151.
36. J.R. Scherer, M.K. Go and S. Kint, J. Phys. Chem. 78, 1304 (1974).
37. S.A. Rice, W.G. Madden, R. McGraw, M.G. Sceats, and M.S. Berger, J. Chem. Phys. in press.
38. B. Curnutte and D. Williams, in Structure of Water and Aqueous Soln, W.A.B. Luck, ed. ISBN 3-527-25588-5 (1974).
39. J.G. Kirkwood, in "Theory of Liquids", B.J. Alder, Ed., Gordon and Breach, New York, N.Y. 1968.
40. C. Tanford, "The Hydrophobic Effect", John Wiley and Sons, New York, N.Y. 1973.
41. F. Franks in "Water-A Comprehensive Treatise", II, F. Franks ed. Plenum Press, New York, N.Y. 1972, p.1.
42. A Ben-Naim in "Water and Aqueous Solution," R.A. Horne, ed., Wiley-Interscience, New York, N.Y. 1972, p. 425.
43. D.D. Eley, Trans. Faraday Soc., 35,1281 (1939).
44. H.S. Frank and M.W. Evans, J. Chem. Phys., 13, 507 (1945).
45. H.S. Frank and W.Y. Wen, Disc. Faraday Soc. 24, 133 (1957).
46. D.N. Glew, J. Am. Chem. Soc., 66, 605 (1962).
47. W. Kauzmann, Adv. Protein Chem., 14, 1 (1959).
48. G. Nemethy and H.A. Scheraga, J. Chem. Phys. 36, 3382, 3401 (1962).
49. V.G. Dashevsky and G.N. Sarkisov., Mol. Phys., 27, 1271 (1974).
50. S.R. Ungemach and H.F. Schaefer III, J. Am. Chem. Soc., 96, 7898 (1974).
51. S.W. Harrison, S. Swaminathan, D.L. Beveridge and R. Ditchfield Int. J. Quantum Chem., in press.
52. J.A. Pople, J.S. Binkley, and R. Seeger, Int. J. Quantum Chem., Symp. No. 10, 1 (1976).
53. M. Yaacobi and A. Ben-Naim, J. Phys. Chem., 78, 175 (1924); H. Edelnoch and J.C. Osborne, Jr., Advan. Protein Chem., 30, 188 (1976).
54. J.D. Bernal and J. Fowler, J. Chem. Phys., 1, 515 (1933).
55. H.L. Friedman and W.D.T. Dale in "Modern Theoretical Chemistry", Vol IV, Statistical Mechanics, pt A, B.J. Berne, ed., Plenum Press, New York, N.Y. 1977.
56. H. Kistenmacher, H. Popkie and E. Clementi, J. Chem. Phys. 61, 799 (1974).
57. M.R. Mruzik, F.F. Abraham, D.E. Schreiber, and G.M. Pound, J. Chem. Phys., 64, 481 (1976).

58. I.R. McDonald, J.C. Rasaiah Chem. Phys. Lett. 34, 382.(1975).
59. C.L. Briant and J.J. Burton, J. Chem. Phys., 60, 2849 (1974).
60. K. Heinzinger and P.C. Vogel, Z. Naturforsch., 29a,1164 (1974); P.C. Vogel and K. Heinzinger, Z. Naturforsch., 30a, 789 (1975).
61. F.J. Millero in "Water and Aqueous Solution", R.A. Horne, ed., Wiley-Interscience, New York, N.Y. 1972, p. 519.

RECEIVED September 7, 1978.

17

Computer Modeling of Quantum Liquids and Crystals

M. H. KALOS, P. A. WHITLOCK, and D. M. CEPERLEY

Courant Institute of Mathematical Sciences, 251 Mercer Street, New York, NY 10012

There are a number of many-body systems which exhibit quantum effects on a macroscopic scale. These include liquid and crystal states of both He-3 and He-4, the electron gas, and neutron matter which probably constitutes the interior of pulsors. In addition, "nuclear matter" - a hypothetical extensive system of nucleons has been studied for the insight one may gain into the nature of finite nuclei. The theoretical studies of these systems have by now a long history, but are by no means concluded. In the last few years, significant advances have been made. This has come in part from the maturity of and gradual unification of many-body theory, in part from the development and application of powerful new expansion procedures, especially varieties of hypernetted-chain equations (1) and finally to the growing power of computer simulation methods for quantum systems. This article is intended as a review of some recent development in computational methods for extensive quantum systems, and of the relation between results so obtained to the evolution of other theoretical work.

Computational modelling of quantum many-body systems is not especially novel. The early history of Monte Carlo methods included many proposals for the solution of Schrodinger's equation with intended application to the many-body problem. Unfortunately, the computational power available was not adequate to do more than simple exercises. The first work in which a significant contribution to theory was made was that of W. L. McMillan (2) who noted that the general sampling algorithm of Metropolis et al. (3) developed to treat equilibrium chemical systems could be used equally well to obtain variational estimates of the energy of a many-body system when the trial function has a product form. Since then, a large number of similar

0-8412-0463-2/78/47-086-219$05.00/0

calculations have been carried out treating extensive systems of atoms from hydrogen to neon. For a thorough review of these calculations see reference (4). It is likely that calculations of this kind will be even more used in the future since they are well suited for modern minicomputers.

We would like to emphasize here some additional methodological developments and their results. The first is the variational treatment of fully antisymmetrized trial functions (5). The second is the Green's function Monte Carlo algorithm (4, 6) which has, in effect, made possible the numerical integration of the Schrodinger equation.

Fermion Monte Carlo

Consider a Hamiltonian

$$H = -\frac{\hbar^2}{2m}\sum_i \nabla_i^2 + \sum_{i<j} v(r_{ij}) \quad . \tag{1}$$

Let R stand for the coordinates of all particles and $\psi_T(R,A)$ be a trial function that depends upon a set of parameters A. Then the variational principle states that

$$E_T \equiv \int \psi_T^*(R) H\psi_T(R)\,dR \Big/ \int |\psi_T(R)|^2 dR \geq E_0 \ , \tag{2}$$

E_0 being the energy of the ground state of the system. McMillan noted that if Eq. (2) were rewritten as

$$E_T = \int |\psi_T(R)|^2 \psi_T(R)^{-1} H\psi_T(R)\,dR \Big/ \int |\psi_T(R)|^2 dR \tag{3}$$

and if a population of points $\{R_m\}$ were drawn at random from a probability density function

$$p_T(R) = |\psi_T(R)|^2 \Big/ \int |\psi_T(R)|^2 dR \tag{4}$$

Then the average over the population of the quantity

$$E_M = \frac{1}{M}\sum_{m=1}^{M} [H\psi_T(R_m)]/\psi_T(R_m) \tag{5}$$

is E_T. Furthermore if we take $\psi_T(R)$ to have the product form:

$$\psi_T(R,A) = \prod_{i<j} f(r_{ij},A) \equiv \exp\{-\frac{1}{2}\sum_{i<j} u(r_{ij},A)\} \tag{6}$$

then sampling p_T is exactly analagous to sampling the Boltzmann distribution since $u(r_{ij},A)$ is a repulsive function which serves to keep the particles apart similar to the role of the potential, $v(r_{ij})$, for classical systems. The sampling may be accomplished by the Monte Carlo method of Metropolis et al. (3). This method is a useful simulation method for Bose liquids like He-4. Crystals can also be studied with a trial function of the form

$$\psi_T(R) = \exp\{-\frac{1}{2}\sum_{i<j} u(r_{ij})\} \prod_{\ell} \phi(r_\ell - s_\ell) \qquad (7)$$

where $\phi(r)$ is a single particle orbital which localizes particle ℓ close to lattice site s_ℓ. Usually ϕ is set as

$$\phi(r) = \exp[-\frac{1}{2} Cr^2] \qquad (8)$$

As with other variational calculations, E_T is minimized with respect to parameters A and C.

The calculation of properties of fermion systems was accomplished by simulating the corresponding Bose system and estimating the effect of antisymmetry by an expansion procedure due to Wu and Feenberg (7) and extended by Schiff and Verlet (8). In the latter, it is assumed that a fermion wave function is

$$\psi_F(R) = \prod_{i<j} \exp[-\frac{1}{2}\sum_{i<j} u(r_{ij})]\det\{\Phi_\ell(r_k,\sigma_k)\} \quad (9)$$

where Φ_ℓ are plane wave orbitals times spin eigenfunctions. Then one expands the energy expectation in successive permutations deriving from the determinant. The first order of correction comes from pair permutations. Schiff and Verlet minimized the corrected fermion energy of liquid states of He-3 with respect to parameters in u, finding apparently good convergence of the Wu-Feenberg expansion. These calculations used a Lennard-Jones potential:

$$v(r) = 4\varepsilon[(\sigma/r)^{12} - (\sigma/r)^6]; \; \varepsilon = 10.22^\circ K, \; \sigma = 2.556 A . \qquad (10)$$

Ceperley, Chester, and Kalos (5) showed that with a trial function of the form given by Eq. (9), the Metropolis algorithm could be carried out directly to sample $|\psi_F|^2$ in a way that was computationally economical in spite of the necessity of evaluating several determinants at each step of the random walk. They found that the convergence of the permutation expansion appears satisfactory at only one density - that for which the pressure is zero. At other densities, or for the kinetic or potential energies separately, the convergence is poor. For example at $\rho\sigma^3 = 0.414$, the fermion Monte Carlo gives an energy of 2.84°K per atom while the first order permutation expansion gives 1.77°K. In fact the first order gives only half the full antisymmetrization correction. It should be stressed that these calculations still do not give correctly the observed energy of liquid He-3 (-1.3°K instead of -2.5°K; of course there is considerable cancellation; the potential energy is about -11°K per atom). This can be ascribed to an incorrect potential but there is even more serious doubt about the accuracy of the product trial function. We will discuss in the next section some consequences of the neglect of three body correlations in the trial function.

Reference (5) also treated fermion systems which model neutron and nuclear matter interacting by simplified pair potentials having the form of a linear combination of Yukawa functions:

$$v(r) = \frac{1}{r} \sum_i \varepsilon_i \exp(-\mu_i r) \quad . \qquad (11)$$

The Wu-Feenberg expansion always underestimated the energy calculated by the fermion Monte Carlo method. In treating crystal phases of fermion systems it is known that the effect of antisymmetry (the exchange energy) is very small. Eqs. (7) and (8) may then be used for the trial functions. The equation of state of fermions interacting with a single repulsive Yukawa was determined for both liquid and crystal phases and a critical strength determined for the existence of a phase transition.

It is also worth mentioning that recently developed integral equations (9, 10) that extend the HNC method to include antisymmetrization (the FHNC equations) give a good account of the neutron fluid energies at low densities. At high densities, different expressions for the kinetic energy that would be equal in a correct variational calculation give rather different answers indicating that the expansions are not well behaved. Interestingly, one of the expressions for the kinetic energy, that due to Pandharipande and Bethe (1), gives good agreement with the fermion Monte Carlo results in all cases. This fact is not yet explained.

Ceperley (11) has applied the methods discussed here to the treatment of an electron gas in both two and three dimensions with a uniform neutralizing background. He considered three possible states: fluid with half the spins up; fluid with all spins aligned;

and crystal phases. He found that at low density the crystal phase is favored. At intermediate densities the totally polarized fluid is most stable and at high densities the equilibrium state is that of the unpolarized fluid. In three dimensions the transition densities occur at $(5.4\pm0.4) \times 10^{13}/cm^3$ and at $(9.2\pm1.8) \times 10^{19}/cm^3$. He finds good agreement for the unpolarized fluid with other theoretical work (12, 13) at low densities but not at high density where the Monte Carlo is at its most accurate. The equation of state for the crystal is in agreement with an anharmonic expansion method (14).

We believe that fermion Monte Carlo will find significant future applications, possibly in Quantum Chemistry.

Green's Function Monte Carlo

This is a class of algorithms which makes feasible on contemporary computers an exact Monte Carlo solution of the Schrodinger equation. It is exact in the sense that as the number of steps of the random walk becomes large the computed energy tends toward the ground state energy of a finite system of bosons. It shares with all Monte Carlo calculations the problem of statistical errors and (sometimes) bias. In the simulations of extensive systems, in addition, there is the approximation of a uniform fluid by a finite portion with (say) periodic boundary conditions. The latter approximation appears to be less serious in quantum calculations than in corresponding classical ones.

We can give here only a sketch of the basis of the method; for more details consult reference (4) and (6).

In dimensionless form and in the space R^{3N} for an N body system, Schrodinger's equation may be written as

$$-[\nabla^2+V(R)]\psi(R) = E\psi(R) \tag{12}$$

where V(R) is the total potential energy.

Suppose the potential energy is bounded from below

$$V(R) \geq - V_0. \tag{13}$$

Then Eq. (12) may be rewritten as

$$(-\nabla^2+V(R)+V_0)\psi(R) = (E+V_0)\psi(R) \tag{14}$$

Consider Green's function for the operator on the left, which satisfies

$$(-\nabla^2+V(R)+V_0)G(R,R_0) = \delta(R-R_0) \qquad (15)$$

and all appropriate boundary conditions for $\psi(R)$. For example, in simulating an extensive system using a finite number of particles and periodic boundary conditions, G must be periodic with respect to every coordinate. Equation 13 implies that $G(R,R_0)$ is always nonnegative and can be interpreted as a density function.

With the help of such a Green's function, Eq. (14) may be formally integrated to give

$$\psi(R) = (E+V_0) \int G(R,R')\psi(R')dR' \quad . \qquad (16)$$

This equation may be solved by iteration giving a function $\psi^{(n)}$ at the nth iterate. Asymptotically $\psi^{(n)}$ is proportional to ψ_0, the ground state wave function. The coefficient is asymptotically constant if E in Eq. (16) is set equal to the ground state energy E_0. It is not difficult to see that if at any stage n one has a population of points $\{R_k^{(n)}\}$ drawn at random from $\psi^{(n)}$, and if one samples $\{R_\ell^{(n+1)}\}$ at random using the density function $(E+V_0)G(R_\ell^{(n+1)},R_k^{(n)})$, then the expected density of the new population near R is

$$\int (E+V_0)\psi^{(n)}(R')dR' \equiv \psi^{(n+1)}(R) \qquad (17)$$

This defines one step of a random walk whose asymptotic density is $\psi_0(R)$.

$G(R,R_0)$ is not known explicitly (or by quadrature) for any but the most simple (and uninteresting problems). But it is clearly related to the solution of a diffusion problem for a particle starting at R_0 in a 3N dimension space and subject to absorption probability $V(R) + V_0$ per unit time. We therefore expect to be able to sample points R from $G(R,R_0)$ conditional on R_0. It turns out to be possible by means of a recursive random walk in which each step is drawn from a known Green's function for a simple subdomain of the full space for the wavefunction. References (4) and (6) contain a thorough discussion of this essential technical point, and also of the methods which permit the accurate computation of the energy and other quantum expectations such as the structure function, momentum density, Bose-Einstein condensate fraction, and crystal structure. One technical point is well worth reiterating here: if one modifies Eq. (16) by multiplying through by $\psi_T(R)$, a trial function of the kind used in variational methods, then the Monte Carlo variance of all quantities may be considerably reduced. When this transformation of Eq. (16) is carried out, it is possible to establish an estimator for the energy whose variance vanishes in the limit $\psi_T \to \psi_0$. In practice we find that optimum variational wavefunctions always reduce the variance of the energy by large ratios. Limited exploration has also shown that significant departures from such trial functions may change the variance (usually, but not always, for the worse) but that within statistics the answers agree.

Results Obtained with GFMC

A number of systems have been studied with algorithms of the class described here. All those to be discussed here used the acceleration technique outlined at the end of the previous section by employing a ψ_T which had the form of Eq. (6) for fluids and Eqs. (7, 8) for crystals.

Ceperley, Chester, and Kalos carried out two studies (15, 16) of boson systems in which pair Yukawa forces (one term with $\varepsilon > 0$ in Eq. (11)) were used. For the most part, the energy values agreed rather well with variational results, usually lying only 1% lower. In certain cases disagreements of up to 4% were noted. On the other hand, the radial distribution function was found to be significantly sharper in the GFMC calculations: the first peak of the radial distribution was usually 10% more peaked than in the corresponding variational calculation.

The generally good agreement with the variational energies indicates that the equations of state of fluids and crystals with Yukawa and similar forces are given adequately for most purposes by the ordinary variational method. For the strongly coupled electron fermion fluid (i.e. at low densities) one can estimate that the change in energy from the corresponding Bose fluid is small. This suggests that the error in using a product trial function as in Eq. (9) is correspondingly small. But there is one point about the comparison of GFMC to variational results that seems generally applicable and important: variational results for the crystal energies are closer to the exact numerical results than are corresponding results for the fluid. That means that the estimation of melting and freezing densities from variational results will have a systematic bias favoring crystal states.

Kalos, Levesque, and Verlet (6) reported a study on the hard-sphere fluid and crystal. They found the energy about 3-5% deeper than had been obtained variationally (17) and a structure function some 10-20% sharper. The authors also developed a perturbation theory connecting hard-sphere and other strongly repulsive potentials. Using this relation they estimated a minimum energy for fluid He-4 with Lennard-Jones potentials (Eq. (10)) as $-6.8\pm0.2^{\circ}$K/atom occuring at $1.0\pm.1$ of the experimentally observed density, ρ_0.

Currently, Whitlock et al. (18) have been treating the Lennard-Jones fluid and FCC crystal by means of the GFMC method. The analysis and calculation of certain corrections is not entirely complete but preliminary results support the perturbation theory results. The equilibrium fluid is found at $\rho/\rho_0 = 1.03$ with an energy of -6.85°K/atom. This is in striking contrast with the variational treatment based on a product wavefunction, Eq. (6) for which no calculation has given a result deeper than -6°K. The experimental result is -7.14°K. Thus the Lennard-Jones parameters of Eq. (10) give substantially better equation of state than had been supposed on the basis of variational calculations. Variational calculations with the product trial function give rather crude upper bounds to the equation of state and hence rather limited information about the He-He potential.

Two theoretical studies have shed light on the discrepancy between the variational and GFMC energies for liquid He-4. Chang and Campbell (19) estimated by a perturbational theory that about half the discrepancy between the two results could be ascribed to the neglect of three-body correlations in the trial function. More recently, Pandharipande (20) used a trial function with a three-body correlation corresponding to a linearized "back-flow" (21). Within the framework of integral equations of the HNC type, he calculated a minimum energy of $(-6.72\pm0.2)^{\circ}$K at a ρ/ρ_0 of 1.05. These three-body correlation effects are undoubtedly important in He-3 as well.

A significant part of the discrepancy between the structure function deduced from variational calculations and from experiment can also be ascribed to the neglect of three-body correlations in the trial functions. Figure 1 shows a comparison of experimental, variational, and GFMC estimates of S(k). It is clear that the latter agrees better with experiment than the

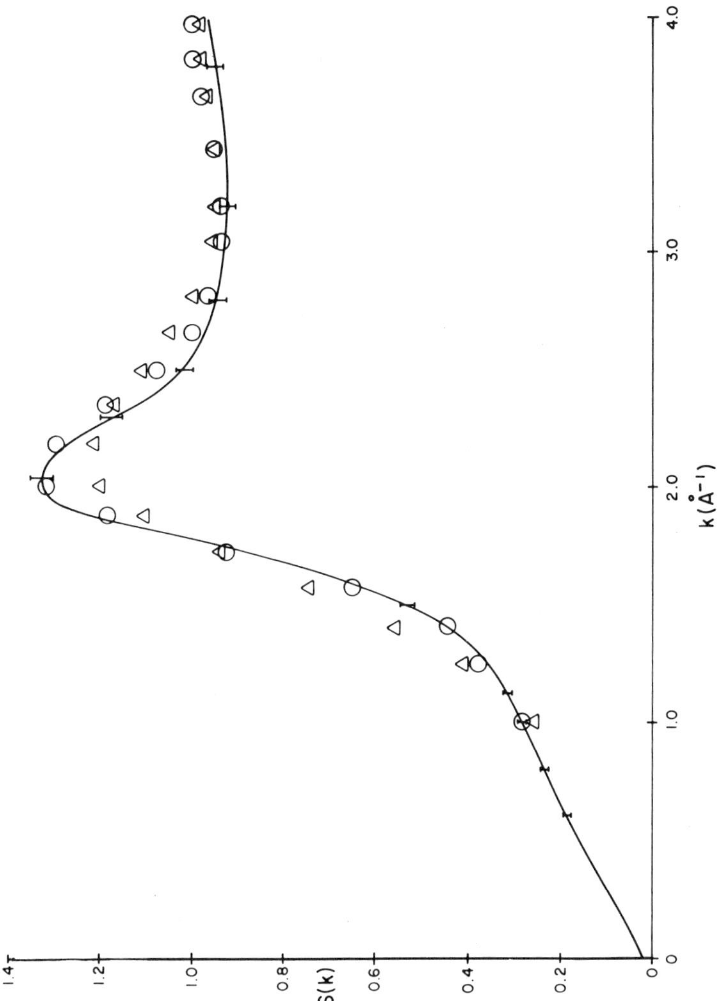

Figure 1. Comparison of structure functions for He-4 at equilibrium density. The solid line shows the smoothed experimental data of Achter and Meyers (Phys. Rev. (1969) 188, 291) *with bars indicating one standard deviation.* S(K) *computed variationally is shown by triangles; the circles show the results computed using* GFMC.

variational result.

It is also clear, in spite of the improvement of equation of state and of S(k) that results from the use of GFMC, that the Lennard-Jones potential cannot be the correct one. Three body potential effects are small in He-4; we estimate that at $\rho = \rho_0$ the triple-dipole force contributes about 0.16^{o}K/atom, fairly close to the result given by Murphy and Barker (22). There is good reason to believe in general that the effect of three body forces on the equation of state will be small as suggested by this particular result. This is a consequence of the high energy of the first excited state of helium. It is not unreasonable to hope that two body forces determined from scattering, virial and transport data should also be consistent with the properties of helium liquids and crystals.

Conclusions

In this brief review we have chosen to concentrate upon the character of some new methods for the Monte Carlo modelling of quantum systems. In so doing we have emphasized certain deficiencies of the older method which rests upon the product trial function in a variational expression. It is necessary to remark that this latter technique remains useful: it is a reasonable guide to the phenomena in quantum systems and for soft-core systems gives results for the equation of state of liquids and crystals which are adequate for most purposes. The extension of the Monte Carlo variational method to include three-body correlations is straightforward but computationally slow; it should be done to provide reliable checks on the theoretical work on such effects in He-4. The study of inhomogeneous systems and mixtures remains largely unexplored.

The fermion and Green's Function Monte Carlo are important in themselves and interesting as hints to the richness of methodology that can be brought to bear on the computation of quantum systems. In the short term we expect to broaden the specific trial functions used in fermion Monte Carlo to permit the more accurate study of He-3 and the treatment of more realistic models of nuclear and neutron matter. We expect also to try a variant in Quantum Chemistry problems.

The most immediate research we plan with the Green's Function Monte Carlo is the exploration of the equation of state of He-4 liquids and crystals with two body potentials which fit more data than the

Lennard-Jones-de Boer-Michels form we have used. The exploration of inhomogeneous systems is a logical next step. At the moment only GFMC offers an unambiguous approach to the calculation of the density profile at an interface, for example.

For the future we anticipate the development of practical methods for the treatment of systems at finite temperatures (some of the technical problems of extending GFMC to temperatures greater than zero have recently been solved for the two-body hard sphere problem (23)). An extension to permit the exact or very reliable treatment of fermion systems seems possible and useful.

Acknowledgment

This work was supported in part by the Department of Energy, Contract No. EY-76-C-02-3077*000 and in part by the National Science Foundation under Grant DMR-74-23494 through the Material Science Center, Cornell University.

Literature Cited

1 Pandharipande, V.R. and Bethe, H.A., Phys. Rev. C, (1973), 7, 1312.
2 McMillan, W.L., Phys. Rev., (1965), 138 A, 442.
3 Metropolis, N., Rosenbluth, A.W., Rosenbluth, M.N., Teller, A.H., and Teller, E., J. Chem. Phys., (1953) 21, 1087.
4 Ceperley, D.M. and Kalos, M.H., "Quantum Many-Body Problems", Chap. IV of Monte Carlo Methods in Statistical Physics, K. Binder, ed.; in press, Springer-Verlag.
5 Ceperley, D.M., Chester, G.V., and Kalos, M.H., Phys. Rev. B, (1977), 16, 3081.
6 Kalos, M.H., Levesque, D. and Verlet, L., Phys. Rev. A., (1974), 9, 2178.
7 Wu, F.Y. and Feenberg, E., Phys. Rev., (1962), 128, 943.
8 Schiff, D. and Verlet, L., Phys. Rev., (1967), 160, 208.
9 Zabolitzky, J., Phys. Rev. A., (1977), 16, 1258.
10 Day, B., Rev. Mod. Phys., in press.
11 Ceperley, D.M., to be published.
12 Stevens, F.A. and Pokrant, M.A., Phys. Rev. A, (1973), 8, 990.
13 Freeman, D.L., Phys. Rev. B, (1977), 15, 5513.
14 Carr, W.J., Phys. Rev., (1961), 122, 1437.
15 Ceperley, D.M., Chester, G.V. and Kalos, M.H.,

Phys. Rev. D, in press.
16 Ceperley, D.M., Chester, G.V. and Kalos, M.H., Phys. Rev. D, (1976), 13, 3208.
17 Hansen, J.-P., Levesque, D. and Schiff, D., Phys. Rev. A, (1971), 3, 776.
18 Whitlock, P.A., Kalos, M.H., Chester, G.V. and Ceperley, D.M., to be published.
19 Chang, C.C. and Campbell, C.E., Phys. Rev. B, (1977), 15, 4238.
20 Pandharipande,V.R., University of Illinois preprint, December, (1977), ILL-(TH)-77-41.
21 Feynman, R.P. and Cohen, M., Phys. Rev., (1956), 102, 1189.
22 Murphy, R.D. and Barker, J.A., Phys. Rev. A, (1971), 3, 1037.
23 Whitlock, P.A. and Kalos, M.H., submitted to J. Comp. Phys., (1978).

RECEIVED September 7, 1978.

18

From Microphysics to Macrochemistry via Discrete Simulations

JACK S. TURNER

Center for Statistical Mechanics and Thermodynamics, University of Texas, Austin, TX 78712

Historically there are two distinct classes of problems in chemistry to which discrete microscopic simulations have been applied widely and with considerable success: At one extreme the bulk physical properties of atomic and molecular fluids are studied as the "exact" dynamical evolution of a collection of representative particles is followed in "computer experiments" using the well-established method of molecular dynamics (1-10). At the other extreme similar techniques (11-16) are used to explore the chemical transformations which may occur in an isolated collision between potentially reactive species (e.g., in a very dilute gas). Between these mutually exclusive limits lies an increasingly important area of chemistry in which macroscopic properties are a direct consequence of cooperative interplay between molecular motion and chemical dynamics. Of immediate interest in this area are chemical systems which may undergo nonequilibrium instabilities and transitions between different regimes of macroscopic physicochemical behavior (See Ref. 17 for a recent survey of the field). Fully analogous to equilibrium phase transitions and critical phenomena, these "nonequilibrium phase transitions" present even more formidable difficulties in both microscopic theory and laboratory experiment. Hence an obvious need exists for detailed computer simulations to provide "experimental" data for theorists and "theoretical" guidance for experimentalists. Since the development of appropriate simulation methods in this area is in its infancy, the purpose of this paper will be to provide motivation for such investigations in the more familiar context of molecular dynamics in classical fluids, to review feasible methods at two levels of description which have emerged in the last few years (18, 19, 20, 21, 22), and to illustrate their application using two simple models which exhibit chemical instabilities and transitions.

In all three classes of problems the systems of interest are characterized by the interaction of large numbers of individual degrees of freedom. It is this feature which leads to great theoretical difficulties in both classical and quantum systems,

0-8412-0463-2/78/47-086-231$08.25/0

and hence provides strong motivation for numerical "experiments" using high-speed digital computers. Appropriate simulations involve molecular dynamics [MD] (1-10) or Monte Carlo [MC] (7, 23, 24, 25) methods which require the details of microscopic interactions and processes. During the course of a simulation this essential information is obtained from a potential energy function (or surface), and it is our limited ability to provide this information which serves to delineate the two extreme problem classes for practical application.

In the first problem class mentioned above (hereinafter called class A), a collection of particles (atoms and/or molecules) is taken to represent a small region of a macroscopic system. In the MD approach, the computer simulation of a laboratory experiment is performed in which the "exact" dynamics of the system is followed as the particles interact according to the laws of classical mechanics. Used extensively to study the bulk physical properties of classical fluids, such MD simulations can yield information about transport processes and the approach to equilibrium (See Ref. 9 for a review) in addition to the equation of state and other properties of the system at thermodynamic equilibrium (7, 8 for example). Current activities in this class of microscopic simulations is well documented in the program of this Symposium. Indeed, the state-of-the-art in theoretical model-building, algorithm development, and computer hardware is reflected in applications to relatively complex systems of atomic, molecular, and even macromolecular constituents. From the practical point of view, simulations of this type are limited to small numbers of particles (hundreds or thousands) with not-too-complicated inter-particle force laws (spherical symmetry and pairwise additivity are typically invoked) for short times (of order 10^{-12} to 10^{-10} second in liquids and dense gases).

In direct contrast to simulations of physical properties dominated by inter-particle forces, the second problem class (class B) concerns interactions among intramolecular and intermolecular degrees of freedom in an event which produces a chemical change in the participants. Severely limited by the need for an accurate quantum mechanical potential energy surface, simulations in this area typically involve classical trajectory (i.e., MD) calculations for the simplest chemical reactions (11-16).

It is not surprising that the two main classes of microscopic simulations have evolved quite independently. Aside from the obvious problem of calculating potential energy functions (surfaces), the greatest computational difficulty arises in treating systems with multiple time scales. Dynamical simulations within class A are feasible because the bulk properties of interest can be determined on a time scale corresponding to a computationally finite number of molecular collisions. When the most important events are rare on this time scale, one rapidly reaches the limits of feasibility for detailed molecular dynamics simulations. With rare exceptions, collisions in which the participants are chemically transformed are indeed rare dynamical events. Hence the combination of potential energy surface and time-scale difficulties means in particular that complex chemical reactions in a many body system are beyond the reach of truly microscopic simulations. That is, the systems of immediate interest in the intermediate class (class C) cannot be treated by straightforward extension (and combination) of class A and class B methods.

This conclusion is certainly nothing new to workers in either of the two main areas of simulation. Indeed much effort in recent years has been directed at precisely the problem which is addressed here in the specific context of chemical instabilities and nonequilibrium phase transitions: How can we simulate the evolution and steady state behavior of systems in which the events of greatest interest are rare on the time scale of individual collisions, but which are nevertheless mediated by the collective many-body dynamics?

Multiple Time Scales, Bottlenecks and Many-Body Dynamics

Systems in which problems of multiple time scales and of "bottlenecks in phase space" occur are hardly exceptional, and promising methods of theory and discrete-event simulation have been developed recently in several important areas. In a marriage of molecular dynamics (and Monte Carlo methods) to transition state theory (26), for example, Bennett (27) has developed a general simulation method for treating arbitrarily infrequent dynamical events (e.g., an enzyme-catalyzed reaction process). An entirely different class of problems has been the recent focus of Adelman, Doll, and coworkers (28, 29, 30, 31, 32), who have investigated the microscopic dynamics of scattering, sorption, and reaction at the gas-surface interface. By taking the many-body lattice dynamics into account in a novel theoretical model based on nonequilibrium statistical mechanics, these authors have developed a new method of molecular dynamics involving generalized Langevin equations of motion.

The new simulation techniques of Bennett (27) and Adelman and Doll (28, 29) illustrate well several important aspects of the present problem. The former considers relatively fast but infrequent events, includes all relevant degrees of freedom expressible in a potential energy surface, and concentrates on the dynamical bottlenecks separating regions of the system's state (phase) space (e.g., products from reactants). In contrast, the latter method treats only a subset of degrees of freedom in detail (e.g., the colliding atom and those surface atoms and neighbors most directly involved), while including in an average way the effect of coupling to other degrees of freedom of the bulk lattice. It is important to realize that both methods treat individual events or sequences of events in great detail, yielding statistical information about a single overall process. Like the class B chemical simulations mentioned above, these methods serve to characterize the "elementary processes" (that is, the reactive events) with which the problems of class C are concerned. In other words, the "fundamental units" of interest in the intermediate class of systems are not simply the atoms or molecules themselves, but rather the reactive collisions and transformations. As we shall see, it is the space-time dynamics and interactions of these new "units" (mediated, of course, by other degrees of freedom via elastic and inelastic collisions, intramolecular dynamics ...) which determine ultimately the macroscopic behavior of such systems. In this respect we see more clearly the connection to the problems of class A. At equilibrium phase transitions in classical fluids, for example, the important "units" are the clusters or nuclei of the new phase embedded in the initial phase. These units are strongly coupled to their surroundings (e.g., via collisions), and in addition are characterized by their own internal "molecular" dynamics.

The time-scale characteristic of size and/or composition changes in individual nuclei is obviously much longer than that of the detailed microscopic dynamics. In particular, the time-delay associated with spontaneous development of a "critical nucleus" has been a source of great computational difficulty in simulations of both liquid→solid (33) and vapor→liquid (34-42) phase transitions in simple classical fluids. Before turning to an analogous situation at the "nonequilibrium phase transitions" which are of immediate interest, therefore, let us first illustrate the key problems and innovations in simulation which are already evident in MD studies of low-order clustering and homogeneous nucleation in the vapor phase (34-42).

Molecular Dynamics in Dilute Gases

In MD simulations of dense fluids, the classical many-body equations of motion can be integrated efficiently using a single-time-step method (e.g., Refs. 3 and 4). This is so because at any instant of time all particles are interacting strongly with a number of others. In a dilute gas, however, only a small fraction of the particles are strongly interacting at any time. If a Lennard-Jones 6-12 (LJ) potential is assumed for a model of fluid Argon, for example, most atoms are at most very weakly interacting via the long-range attractive tail of the potential. With a time-step selected for the few colliding pairs of atoms, the efficiency of the usual MD methods is lost. Recognizing that each atom spent most of its time in essentially free flight between binary collisions, Harrison and Schieve (34, 36) introduced a cutoff (R_∞) in the range of the LJ potential and combined the Alder (1) free-flight method for "isolated" atoms with a Rahman (3) or Verlet (4) continuous-potential method for the remaining atoms. (For details of the dilute gas algorithm see Ref. 36.) The result was a hybrid MD algorithm with which simulations of a dilute classical gas could be performed efficiently in two and three dimensions (34, 35, 36, 37, 38).

Low-Order Clustering in a Computer Model of Argon Vapor. Using the algorithm which decoupled isolated atomic motion from collisions between atom pairs, Harrison and Schieve (34, 35) first studied the approach to equilibrium in a 2D gas of (LJ) Argon. At lower temperatures and higher pressures, low-order clustering was observed, beginning with the formation of bound states (dimers) by the three-body channel (35, 36, 37, 38)

$$\begin{aligned} A + A &\rightleftarrows A_2^* \\ A_2^* + A &\rightarrow A_2 + A \end{aligned} \qquad (1)$$

In the first step an unstable (classical) excited state A_2^* of the dimer is formed which has positive total energy. Enough energy is then removed by a third atom in the second step to form the stable dimer A_2. A typical sequence is shown in Figure 1, in which a dimer 12 is created in a collision with 3 and destroyed after a few vibrations in a collision with 6. Detailed MD studies of the dimer formation "reaction" have been carried out in two and three dimensions (35, 36, 37, 38). The result of greatest interest for the present study is the time-scale problem introduced by the occurrence of the "chemical reaction" (Eq. 1).

Since the dimer-formation process is a dynamically rare event, lengthy computer runs (of order 40,000 collisions in 2.4 x 10^{-7} sec. model time) were required to obtain statistically reliable values for the dimer formation rates, lifetimes, and other properties. The problem to be faced in this situation is a classic one in computer simulation: What level of microscopic detail must be included in order to retain the essential features of the system at a not unreasonable cost? The solution of this problem in the dimer study illustrates well the general approach of this paper: Beginning with the most detailed simulation possible (which then serves as a benchmark), develop a hierarchy of simulations in which the computer model at each level incorporates, in an average way, at least, those features of lower levels relevant to the questions of interest. In this way the qualitative behavior and properties of a system can be mapped with great efficiency, and areas of particular interest so identified could be subjected to more careful, even quantitative, study in lower-level simulations.

For the dimer simulation two simplifications were made: First was the obvious step of restricting the study to two dimensions, at least initially. Second, the continuous LJ potential was replaced by a square-well hard-core (SWHC) potential with parameters determined empirically from equation-of-state data for Argon (cf, Ref. 43, p. 158). With these two refinements it was possible not only to improve the statistics of the dimer calculations, but also to consider still lower temperatures and higher pressures at which trimers, tetramers and other low-order clusters could be formed (39, 40, 41).

Molecular Dynamics Evidence for Vapor-Liquid Nucleation. Specializing the MD model now to a 2D system of 100 SWHC "Argon" atoms, let us review briefly relevant aspects of these simulations. The SWHC potential is defined with well-depth $\varepsilon = 2.305 \times 10^{-14}$ erg, hard core radius $\sigma_1 = 2.98 \times 10^{-8}$ cm, and range of attractive interaction $\sigma_2 = 1.96\ \sigma_1$. The atomic mass is $m = 6.628 \times 10^{-23}$ g. With periodic boundary conditions, the system is adiabatic, at constant volume, total energy, and number of atoms. Initially the atoms are displaced slightly from the sites of a square lattice 3.334×10^{-6} cm ($112.08\ \sigma_1$) on a side, and are given a fixed speed with random velocity orientation. This fixes the total energy of the (initially non-interacting) system to be the initial kinetic energy of the atoms. As clusters are formed (with negative potential energy), therefore, the kinetic energy (i.e., temperature) increases accordingly until an equilibrium distribution of monomers, dimers, trimers, . . . is eventually reached.

A quite remarkable feature of the MD studies of Zurek and Schieve (40, 41, 42) using the above computer model is the direct observation of nucleation in the vapor-liquid phase transition region. The phase transition itself is clearly indicated in Figure 2; the heat capacity $C_V = \partial E(T^*)/\partial T^*$ is seen to increase dramatically near $T^*_{eq} \simeq 0.5$. Here $T^* = kT/\varepsilon$ and the total energy $E^* = E/\varepsilon$ (or T^*_{in}) is plotted against the equilibrium kinetic temperature T^*_{eq}. (Initial conditions with negative total energies are obtained by decreasing the temperature in the final state of higher-energy runs.) In Figure 3 a "supercritical" cluster is clearly evident in a "snapshot" of the system at $T^*_{in} = 0.2$ (41). This qualitative indication that nucleation has occurred is made quantitative in terms of the physical cluster distributions plotted in Figure 4. Here a dramatic change in the distribution is seen as the phase transition region is entered in 4c and 4d. The gap which appears separates clusters into sub- and supercritical classes, and the width of the gap provides bounds on the number of atoms in the "critical" cluster. It is important to realize that nucleation has introduced yet another time scale into the MD simulation--the time required for spontaneous generation of the "critical" nucleus which, in a non-adiabatic system, initiates an irreversible transition from the metastable initial vapor state to the stable liquid phase. Although the implications of this time scale have not been treated in the Argon studies, they will be examined in the simulations of nonequilibrium phase transitions which follow. (Note: These nucleation results have now been verified in a recent MD/MC study by Rao, Berne, and Kalos, Ref. 54.)

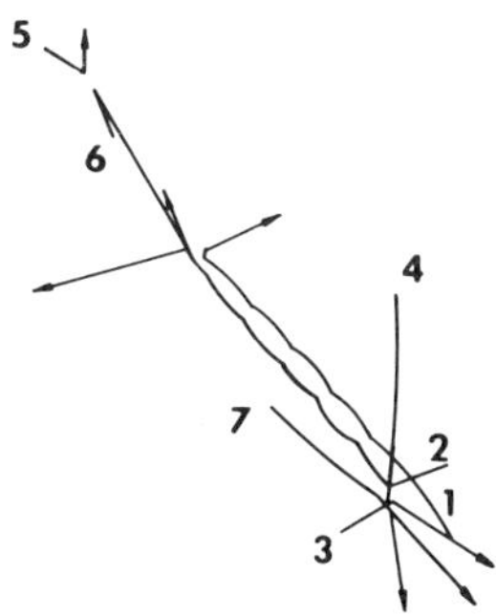

Journal of Chemical Physics

Figure 1. Plot of particle trajectories involved in the formation and destruction of a dimer (12) during a molecular dynamics simulation of 2-D (LJ) Argon (35)

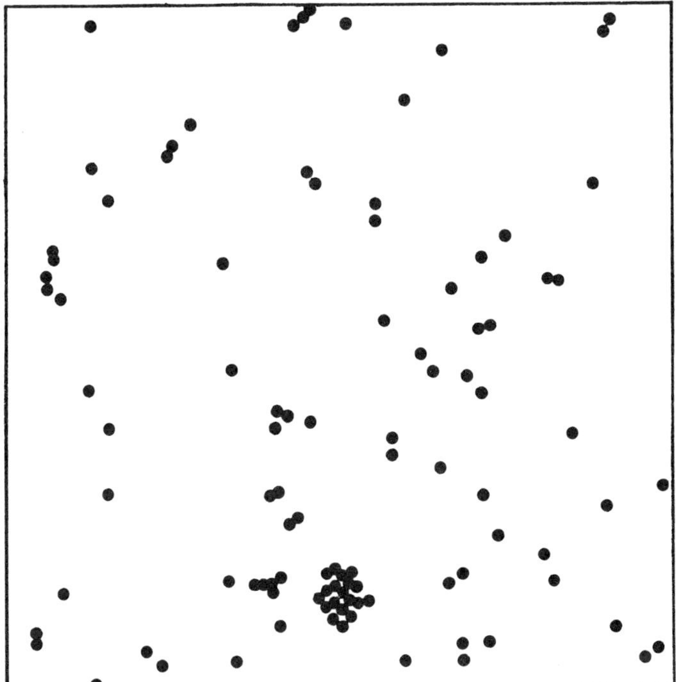

Journal of Chemical Physics

Figure 2. The spatial distribution of the particles in the system at a given instant shows a well-developed droplet containing 19 particles. The size of the points in the picture corresponds to the size of the outer wall of the potential (σ_2). $T^*_{INIT} \cong -0.2$, $T^*_{EQ} \cong 0.5$ *(41).*

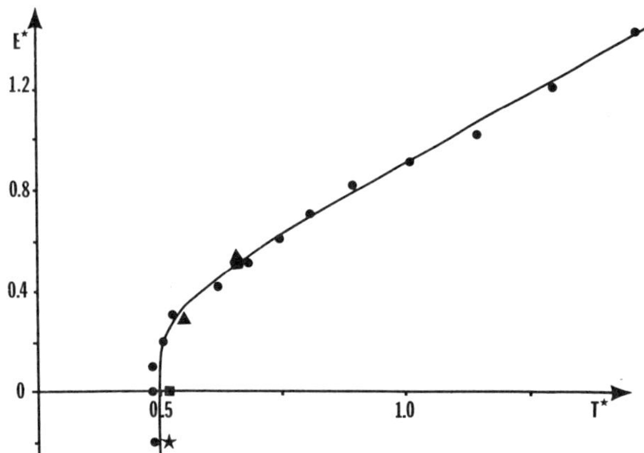

Figure 3. The E(T*) *[*$T^*_{INIT}(T^*_{EQ})$*] dependence, showing the characteristic shape for* $T^* \cong 0.5$. $T^*_{INIT} \leqslant 0$ *was achieved by cooling down the system that has already evolved in the higher* ($T^*_{INIT} > 0$) *temperature for some time. The slope is the specific heat* c_v. *After Ref. 42.*

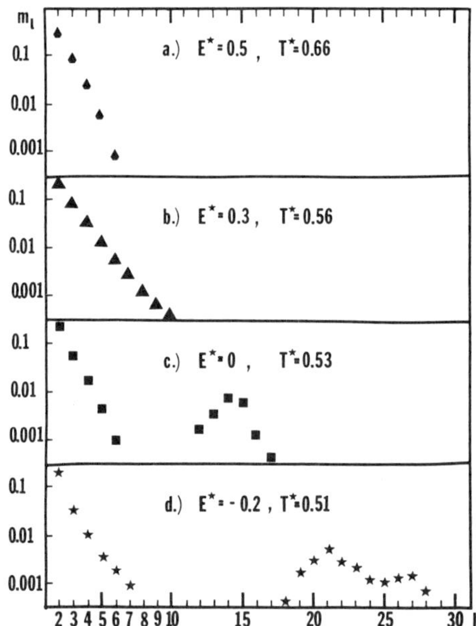

Figure 4. Mole fractions, m_l, *for clusters formed in the adiabatic system at different energies,* E*. *After Ref. 42.*

Chemical Instabilities, Nonequilibrium Phase Transitions, and Dissipative Structures

Let us now digress briefly from matters of simulation to introduce the kinds of chemical systems in which nonequilibrium phase transitions can occur. This is most conveniently done at the macroscopic level of deterministic chemical kinetics. Consider a chemical system under open or closed conditions so that it is possible to maintain a time-independent macroscopic state. At or near thermodynamic equilibrium, the macroscopic description of such a system yields a unique stationary solution that is always stable with respect to arbitrary perturbations. Sufficiently far from equilibrium, however, the governing kinetic equations for nonlinear chemical systems may admit solutions other than the continuous extension of the equilibrium solution into the nonequilibrium domain. Beyond a critical distance from equilibrium, moreover, this thermodynamic branch may become unstable in response to fluctuations occurring in the medium. Beyond the point of instability, the response of the system to an infinitesimal disturbance leads ultimately to a new operating regime characterized by organization in space, time, or function. Whatever the outcome of such a macroscopic instability, therefore, the instability itself originated in the response of the system to a fluctuation. Consequently a purely deterministic description of the system in terms of mean values alone is no longer adequate, and it is essential to supplement the "average" description with a theory of fluctuations extended to nonlinear systems under far-from-equilibrium conditions.

At thermodynamic equilibrium fluctuations constitute a negligible correction to the macroscopic description of matter except near instability points such as phase transitions or critical points. Similarly, in nonequilibrium situations as well one expects that fluctuations become important only near points of nonequilibrium instability (see Ref. 17 for a detailed discussion). For nonlinear systems in which such instabilities and transitions can occur, the role of fluctuations is even more dramatic than at equilibrium, due to the existence in general of many distinct "phases" compatible with the external conditions imposed on the system. This fact has been known for several years now, since the first (numerical) demonstration on simple chemical models of multiple solutions to the governing macroscopic differential equations (44, 45, 46, 47). The variety of possible solutions includes time-independent homogeneous and inhomogeneous states as well as states which are organized in time (homogeneous chemical oscillations) or in time and space (travelling chemical waves). Which of these "dissipative structures" (17) will arise under given conditions depends crucially on the specific type of perturbation to which the original state is initially unstable. In terms of spontaneous fluctuations, which are present in general over a range of wavelengths, both the final state and the evolution toward that state will be determined by the space-time characteristics of the fastest-growing unstable mode.

As one would expect, developments in the theory of such phenomena have employed chemical models chosen more for analytical simplicity than for any connection to actual chemical reactions. Due to the mechanistic complexity of even the simplest laboratory systems of interest in this study, moreover, application of even approximate methods to more realistic situations is a formidable task. At the same time a detailed microscopic approach to any of the simple chemical models, in terms of nonequilibrium statistical mechanics, for example, is also not feasible. As is well known, the method of molecular dynamics discussed in detail already had its origin in a similar situation in the study of classical fluids. Quite recently, the basic MD computer model has been modified to include inelastic or reactive scattering as well as the elastic processes of interest at equilibrium phase transitions (18), and several applications of this "reactive" molecular dynamics (RMD) method to simple chemical models involving chemical instabilities have been reported (18, 19, 22). A variation of the RMD method will be discussed here in an application to a first-order chemical phase transition with many features analogous to those of the vapor-liquid transition treated earlier.

Reactive molecular dynamics may be viewed as a microscopic computer experiment inasmuch as all relevant particle processes can be included to some degree. In particular the method treats elastic as well as inelastic or reactive collisions between particles and, indeed, requires that the latter be relatively rare events in order to simulation actual chemical kinetics in a realistic way. When large enough systems are considered (e.g., thousands of particles) it should be possible to "measure" temporal and spatial correlation functions, for example, and to make quantitative the notion of critical size and amplitude of fluctuations necessary to nucleate a macroscopic transition. A first step in the latter direction will be reported here.

Applications to date of reactive molecular dynamics methods demonstrate feasibility for the study of hundreds or thousands of particles involved in chemical reactions for which reactive probabilities do not vary too widely. The latter condition is essential if statistically significant numbers of all possible events are to occur within a practical computation period. Even if the requirement for a large excess of elastic collisions is relaxed, however, reaction rates typical of experimental chemical systems demand simulation run times which approach the feasible limit except for quite small numbers of particles. Turning therefore to a higher level method for this type of system, one may treat numbers of particles in a small cell or volume element rather than individual particles, and invoke a Monte Carlo procedure for determining which reactive event will occur, how much time will elapse between events, and in which "cell" of the system the process will occur. Transport between adjacent cells is handled in a similar fashion.

In such a method, only the events of greatest interest (reaction or transport) are treated, so that widely different rate constants need not be a problem. In the change from reactive molecular dynamics to stochastic simulation, much larger systems may be considered, and computational efficiency is gained at the expense of molecular detail. In addition, problems related to numerical "stiffness" associated with the deterministic differential equations (48, 49) are completely avoided in a stochastic simulation. Finally, a number of implementations of such techniques have been presented which give qualitatively reasonable results. Of particular significance is a recent formulation due to Gillespie (20), in which the transition probabilities and Monte Carlo implementation are rigorous consequences of the very assumptions from which chemical master equations of the birth-and-death type are derived. This fact is of great importance for the development and testing of theory. Specifically, computer experiments based on the latter formulation simulate directly a stochastic chemical evolution satisfying a master equation of the type which is central to theoretical developments in this area (17). A report of initial investigations of chemical instabilities using stochastic simulation methods will also be presented.

In order to fix the main ideas most efficiently, the microscopic chemical simulations will be presented in applications to simple chemical models which together exhibit most of the known types of instabilities and transitions. As will be seen, both RMD and stochastic simulations treat the elementary collision and transport processes which occur in a chemical mixture instead of the concentration (or populations) of the constituents themselves. In the language of a Markovian stochastic description of chemical kinetics (cf, Ref. 17), both of these methods may be expressed in terms of the probability $P(\tau,\mu)\delta\tau$ that the next event will occur in the infinitesimal time increment $\delta\tau$ following an interval τ and will be an event of type μ. Such a probability leads naturally to a procedure for simulating a system's time-evolution in a "computer experiment." Each method is distinguished therefore by an algorithm for evaluating $P(\tau,\mu)$ and for implementing an iterative procedure to simulate an evolution of a chemical system.

Microscopic Chemical Simulations: Reactive Molecular Dynamics

In designing a simulation or "computer experiment" in a particular context, one must first determine the level of detail needed to describe the phenomena of interest. For chemical reactions, in which the fundamental interactions are between atoms and molecules, one would expect to begin at the level of microscopic dynamics. Simulations of this kind follow the details of classical many-body dynamics derived from assumed intermolecular interactions, and constitute the by now familiar method of molecular dynamics. As we have seen, such microscopic simulations of chemical reactions are impractical at present, largely due to the enormous number of "uninteresting" dynamical events which must be followed in detail between the reactive collisions of greatest concern. One way of avoiding part of the difficulty is to adopt a molecular "pseudodynamics" approach in which molecules travel in free flight between instantaneous hard-sphere collisions at which reactive processes may occur. Such methods have been employed recently by Portnow (18), by Ortoleva and Yip (19), and by the present author (22) and are applied here to some questions raised in the deterministic and stochastic formulations of chemical instabilities.

First Order Chemical Phase Transition in a Cooperative Isomerization Reaction. A convenient model of a first order transition is provided by a reversible isomerization reaction in a macroscopically homogeneous system

$$A \underset{k_2}{\overset{k_1}{\rightleftarrows}} B \tag{2}$$

having activation energies ε for the forward reaction and η/N per molecule of B present in the system for the reverse reaction, where N is the total number of molecules. The activation energy for the reverse reaction depends on the number N_B of B molecules present, and therefore introduces a cooperativity into the mechanism. In physical terms, this cooperativity expresses the stabilizing influence on a B isomer of intermolecular interactions with other isomers of the same type. Aside from being a convenient way to introduce nonlinearity into the simplest chemical model, this procedure yields a model which is isomorphic to a Bragg-Williams (mean field) model for monomolecular adsorption on a uniform surface. The latter model of the equilibrium two-dimensional lattice gas is known to exhibit a first-order phase transition (cf, Ref. 50) and provides a convenient vehicle for an especially intuitive discussion of the stochastic theory of metastability and nucleation at chemical first-order transitions (22, 51). (The use here of this equilibrium example is motivated by practical considerations of theory and simulation, and does not limit the generality of the results and conclusions.)

If the cooperativity is introduced into the rate constant for the reverse reaction via a "mean-field" dependence on N_B, then the equilibrium constant for the isomerization reaction takes the form

$$K = \exp\,(-\varepsilon + \eta N_B/N) \tag{3}$$

where the activation energies are expressed in units of k_BT, with k_B the Boltzmann constant and the T the absolute temperature. If N is fixed (closed system), then the kinetics of the reaction is given by a single macroscopic rate equation

$$\frac{dN_B}{dt} = k_0(N - N_B)e^{-\varepsilon} - k_0 N_B e^{-\eta N_B/N} \qquad (4)$$

Defining a new time variable $\tau \equiv k_0 e^{-\varepsilon} t$ and the mole fraction $x = x_B \equiv N_B/N$ gives

$$\frac{dx}{d\tau} = 1 - x - xe^{\varepsilon-\eta x} \qquad (5)$$

A linear stability analysis of the steady state x_0 of this system implies stability whenever the real normal mode frequency ω is negative, and instability otherwise, where $x\omega = -1 + \eta x - \eta x^2$. Points of marginal stability correspond to $\omega = 0$ and, according to the quadratic expression for ω, there may be 0, 1, or 2 points. It is easy to see that there is no instability for $\eta < \eta_C = 4$. For $\eta > \eta_C$, however, there is a region of ε in which three steady states are possible. Typical steady state curves are shown in Figure 5. Here the curve passing through the critical point $(\eta_C, \varepsilon_C, x_C) = (4, 2, 1/2)$ separates regions of 1 and 3 steady states. In the three state region, the upper and lower branches are stable, the middle branch unstable, according to linear stability analysis.

If it is supposed for convenience that ε is an externally variable parameter of the model, then for $\eta > \eta_C$ the transition between stable branches of the deterministic solution should occur discontinuously at the points of marginal stability (arrows, Figure 5). The similarity between this picture and the pressure-volume diagram for the equilibrium liquid-vapor transition is apparent. Indeed, according to the macroscopic van der Waals model of a nonideal gas (50) the liquid-vapor phase change in a homogeneous fluid should occur at a lower pressure than the reverse transition. That a single equilibrium transition pressure (e.g., dashed line, Figure 5, an "equal-areas" construction) is observed regardless of the direction of change is a direct consequence of the response to spatially localized, finite-amplitude fluctuations in the initial fluid state. Moreover, when such fluctuations are of insufficient size or amplitude to provide a stable nucleus for the transition, then states beyond the equilibrium transition point may be realized. Such states are termed metastable, since they are stable only until a large enough fluctuation appears to form a "supercritical" nucleus. At which point along the metastable branch will destabilizing fluctuations appear, and how long will the metastable state remain "stable" prior to occurrence of critical fluctuations? Clearly such questions are beyond the scope of the deterministic description, and will be addressed here from the viewpoint of RMD simulation.

Reactive Molecular Dynamics. A microscopic computer model of a chemically reacting mixture is constructed with the following essential features: (1) Hard-sphere dynamics. The key assumption is that the most important chemical effects occur upon energy and momentum transfer at molecular collisions, the net effect of attractive intermolecular forces being to alter the collision cross-section. Hence the microscopic molecular dynamics is simplified enormously by replacing a realistic intermolecular potential by a hard core repulsion. The molecules are then approximated by rigid spheres, and in the present model are assumed in addition to have no internal degrees of freedom. As in the MD model of a dilute gas discussed earlier (34, 36), the molecular dynamics reduces then to free-flight motion between collisions, and results in a computational efficiency which makes it possible to include chemistry at all. (2) Reactive transitions. Chemistry is introduced into the model by assigning reaction probabilities to members of a colliding pair at the instant of collision. This process is handled conveniently, and in a theoretically consistent fashion, by assuming an activation energy for each possible reaction channel, and selecting in a prescribed way which channel is to be followed. If a single reaction is possible, for example, the condition for reaction is simply that the relative kinetic energy of the colliding pair exceed the activation energy for the process. Once a particular reaction channel is selected, molecular identities and other properties of the reacting molecules are altered according to the laws governing the particular chemical transformation. All particles then move in free flight until the next collision occurs and the channel-selection process is repeated.

It is easy to see that these two components of the RMD simulation provide the information contained in the reaction probability function $P(\tau,\mu)$. The molecular dynamics determines the type of collision C_k and the elapsed time τ, while the probability of reactive process R_μ occurring upon collision C_k is given independently in the second step.

A Discrete Model of Cooperative Isomerization. The algorithm which implements the method of reactive molecular dynamics (RMD) is understood best in the context of a specific application. Therefore let us specialize now to the simple isomerization reaction (Eq. 2) to address the questions of "chemical nucleation" raised in that context. Only the main ideas are stressed here. A more detailed account of these studies appears in Ref. 22.

Consider a binary mixture of constituents A and B, and assume that the reversible isomerization is a collision-induced reaction. There are then four elementary reactive processes which may occur upon binary collision in the system:

$$\text{(a)} \quad A + A \xrightarrow{k_1} A + B \tag{6a}$$

$$\text{(b)}\quad A + B \overset{k_1}{\rightarrow} B + B \qquad (6b)$$

$$\text{(c)}\quad A + B \overset{k_2}{\rightarrow} A + A \qquad (6c)$$

$$\text{(d)}\quad B + B \overset{k_2}{\rightarrow} B + A \qquad (6d)$$

Since the system is closed, the total number $N \equiv N_A + N_B$ is constant, and it is easy to verify that in this case the deterministic rate equation for N_B reduces to the form of Eq. 4, where the constant N is absorbed into the pre-exponential factor k_0 of the latter equation.

The sequence of collisions, and hence a partial determination of reaction channel, is generated by the hard-sphere molecular dynamics algorithm. For collisions of type A-A or B-B, the choice is clear, and it remains to decide whether the condition for reaction, given the appropriate collision, is satisfied. For type A-B collisions there are two possible channels from which to choose. Thus transition probabilities $W_\mu(C_k)$ are assigned which give the probability for reaction R_μ to occur given a collision of type C_k. Suitably normalized, these transition probabilities for the four channels of Eq. 6 are

$$W_{A\rightarrow B}(A\text{-}A) = e^{-\varepsilon}$$

$$W_{A\rightarrow B}(A\text{-}B) = \frac{e^{-\varepsilon}}{1 + e^{-\varepsilon}}, \quad W_{B\rightarrow A}(A\text{-}B) = \frac{e^{-\eta N_B/N}}{1 + e^{-\varepsilon}} \qquad (7)$$

$$W_{B\rightarrow A}(B\text{-}B) = e^{-\eta N_B/N}$$

The probability of no reaction, given collision C_k, is then

$$W'(C_k) = 1 - \sum_{\mu=1}^{M} W_\mu(C_k) \qquad (8)$$

Here it is assumed that at most one reactive event per collision can occur. Which of the possible events R_μ occurs for a given collision is then determined by picking a uniform random number α from the unit interval. In general, therefore, if

$$\sum_{\nu=0}^{\mu-1} W_\nu(C_k) < \alpha < \sum_{\nu=0}^{\mu} W_\nu(C_k)\ , \quad \mu = 1, 2, \ldots;\ W_0 = 0 \qquad (9)$$

then process R_μ is selected. Once a particular channel has been selected, particle identities are altered according to the appropriate reaction path of Eq. 6. For A-A and B-B collisions, a second random number is used to determine which of the (equally likely) molecules will react.

In order to incorporate the macroscopic cooperativity parameter η into the dynamical model, consider what η means at the molecular level. Essentially η represents an average interaction energy, of a B molecule with other molecules of the same type, which tends to stabilize the given molecule with respect to conversion into the isomeric form A. Since intermolecular forces are of relatively short range, η contains the effects of interactions with nearest neighbors, next-nearest neighbors, etc., the strength of which decreases with distance. Hence $\eta N_B/N$ should properly reflect the influence of the local environment on the stability of a B molecule. This localization of the cooperativity is accomplished conveniently by introducing a "mean-field radius," R_{mf}. The instantaneous reaction probability for a colliding B molecule becomes

$$W_{B\to A} \sim \exp\,[-\eta(N_B/N)_{R\le R_{mf}}] \qquad (10)$$

which now includes only the fraction of B molecules within a mean-field radius of the reference B-molecule. In this relation η is identified with the deterministic quantity. In a dynamical simulation, therefore, R_{mf} can be determined by fitting the observed steady state (mole) fraction N_B/N to the deterministic curves (Figure 5). Such an empirical value of R_{mf} will necessarily depend on the parameters which define the rigid sphere molecular dynamics model. If the interpretation given here is valid, however, R_{mf} should not depend on the chemical parameters ε and η or on the total number N of molecules.

In the simulations reported here, several simplifying assumptions have been made which have been relaxed in more recent studies: (a) The system is maintained isothermal by requiring that collisions are dynamically elastic, with no thermic effects due to heats of reaction. (b) The system is two-dimensional, which facilitates greatly visualization of the chemical dynamics and reduces the complexity of the rigid-molecule dynamics. (c) The molecules are physically identical, with mass m and hard-sphere (disk) diameter σ

Implementing the RMD Algorithm. Because the collisions are instantaneous in a rigid particle system, the "time to next collision," t_c, is easily evaluated in terms of the positions and momenta of the N particles at any given time t (7). The particles are moved in straight lines to their new positions at time $t + t_c$, the momenta of the colliding pair are exchanged, and t_c reevaluated. Hence the dynamical evolution of the model N-body system may be followed exactly in a computer experiment. Obviously the computer storage and time requirements increase with N, so it is important to use as small a system as can reasonably simulate the phenomena of interest. We consider 100 rigid molecules in a two-dimensional region of area a and impose periodic boundary conditions to simulate a typical "cell" in a much larger system. Despite this generalization, the total number N of particles within the cell is preserved under periodic boundary conditions. If instead rigid (reflecting) boundaries were imposed, then the behavior of such small systems would be dominated by the walls, especially when spatially localized effects are important. The chemical evolution of the system would depend strongly on the size of the enclosing "box," and any connection to macroscopic chemistry would require extrapolation of small-system results. Undesirable size effects can be minimized, but not eliminated, with periodic boundary conditions. The possibility of boundary-related spatial correlations is then introduced, however, especially for very small "cells"; the significance of such correlations in the present model has yet to be established.

The choice of dynamical initial conditions is not crucial, since both thermalization of velocities and randomization of positions occur typically within a few collisions per particle. Hence a convenient square lattice initial configuration is chosen, with uniform speed $|v| = (2k_BT/m)^{1/2}$, corresponding to the given temperature T, and random direction vector. A "snapshot" of this initial condition for $a = 196\sigma^2$ is shown in the first frame of Figure 6.

Once chemical identities are assigned to the N molecules in the system, an evolution which simulates microscopic chemical kinetics can be followed, and appropriate dynamical and chemical data accumulated. In order to compare results with stochastic or deterministic theory (cf, Ref. 22 in the present case), it is necessary to generate suitable "ensemble" averages by repeating each run a number of times (19, 22). This essential but expensive procedure will be unavoidable whenever the molecular dynamics depends on the chemistry (e.g., when "internal" energy is converted into kinetic energy via an exothermic reaction). A feature of the present computer model, by virtue of assumptions (a) and (c) above, is that particle motion can be decoupled from chemical processes. This means that each set of dynamical data can be generated without regard to chemistry and stored for repeated chemical simulations using different distributions of initial chemical identities, different transition probabilities, or even entirely different chemical mechanisms. The result is an enormous savings in computer time, since the molecular dynamics remains the most time-consuming part of the RMD algorithm despite the use of hard-particle dynamics. Obviously, enough dynamical runs should be used to eliminate undesirable correlations due to repeated use of the space-time sequence of "common random numbers" generated in a single run. Even so, "ensemble" averages based on hundreds

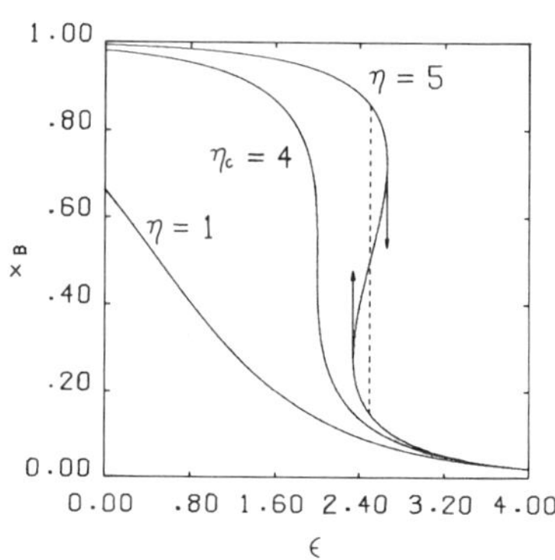

Journal of Physical Chemistry

Figure 5. Steady states of the cooperative isomerization model, showing subcritical ($\eta = 1$), critical ($\eta = 4$), and supercritical ($\eta = 5$) curves of mole fraction x_B as a function of the forward activation energy ϵ. The deterministic transitions for $\eta = 5$ are indicated by arrows; the dashed line denotes an "equal areas" construction which determines the unique equilibrium transition (22).

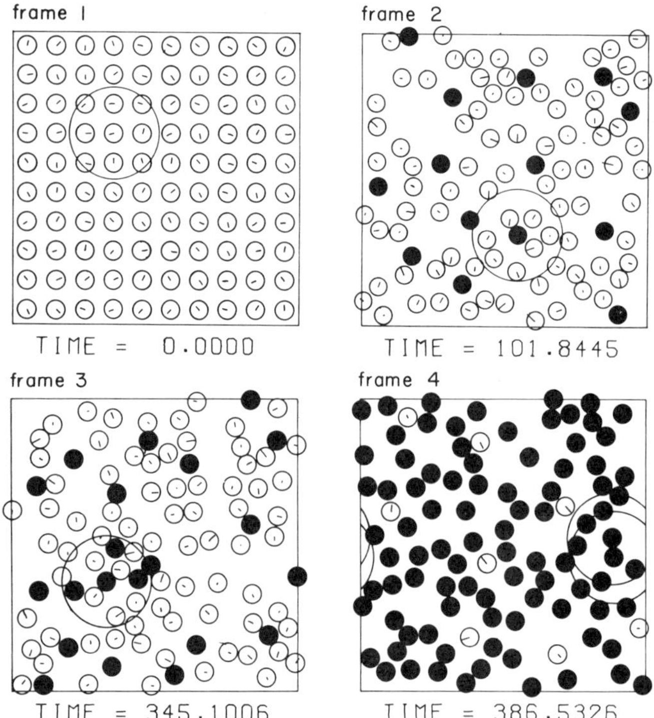

Journal of Physical Chemistry

Figure 6. "Snapshots" of an RMD simulation showing nucleation of the high-B phase, with $\eta = 5$ and $\epsilon = 2.348$, beginning in the pure A state ($x_B = 0$ in Figures 5 and 7). Darkened disks denote B molecules; the short line inside each disk indicates the velocity vector. The mean-field radius $R_{mf} = 2.50\sigma$ is drawn in each frame. In frames 2–4 it is centered at a colliding B molecule to indicate the fluctuation in mean-field cooperativity, especially evident in comparing frames 2 and 3 (22).

of chemical evolutions can be generated easily using common dynamical data from two or three runs and with no significant correlation effects.

RMD Simulation of Chemical Nucleation (22). A series of microscopic computer experiments was performed using the cooperative isomerization model (Eq. 2). This system was selected for the trial simulations for several reasons: First, only two chemical species are involved, so that a minimal number of particles is needed. Second, the absence of buffered chemicals (e.g., A and B in the Trimolecular reaction of the next section) eliminates the need for creation or destruction of particles in order to maintain constant populations (19, 22). Third, the dynamical model of the cooperative mean-field interaction can be examined as a convenient means of introducing cubic or higher nonlinearity into molecular models based on binary collisions. Finally, the need for a microscopic simulation is most apparent for transitions between multiple macroscopic states. Indeed, the characterization of spatially localized fluctuations is of obvious importance to the understanding of nucleation phenomena. As for the equilibrium vapor-liquid and liquid-solid transitions, detailed simulations at the molecular level should provide deep physical insight into chemical nucleation processes which is unattainable from theory, higher-level simulation, or experiment.

The RMD simulations used dynamical data generated by three molecular dynamics runs of 10^5 collisions each. The runs involved 100 identical particles, in a box of side 16σ, with periodic boundary conditions. The dynamical initial conditions were identical except for the assignment of particle velocity directions using a different set of random numbers in each run. If $a_{cp} = (N/2)\sqrt{3}\,\sigma^2$ is the area occupied by N particles at close packing, then the close packing ratio $r \equiv a/a_{cp} = 2.956$ in these runs. Under these conditions the mean free path between collisions is $\ell_r \sim 1.14\,\sigma$ or roughly 1/14 of the box lenth.

It was anticipated that the most sensitive feature of the macroscopic model to the cooperativity would be the location of the critical point $(\eta_c, \varepsilon_c, x_c) = (4,2,.5)$. Hence the fit of simulation results to the critical curve of Figure 5 was used to select the best mean-field radius for the system. The agreement obtained with $R_{mf} = 2.50\,\sigma$ is shown in Figure 7. In this figure the curves are the deterministic results, and each data point represents the average of 11 RMD runs of 10^5 collisions each, beginning with the same initial mole fraction $x \equiv N_B/N$, with different random initial assignment of particle identities.

In the simulations of Figure 7 it is apparent that most of the points studied on the metastable branches for $\eta = 5$ are in fact stable on the time scale of these computations. This should not be surprising, since the "size" of a destabilizing fluctuation is large near the "equal-areas" construction (dashed line, Figures 5 and 7), and decreases to microscopic dimensions only at the marginal stability points of the deterministic curve (22, 51). Such large fluctuations should be extremely rare, and in particular require induction times much longer than the 10^{-9} sec duration of the 10^5 collision sequence followed in these simulations. How then can one expect to see nucleation occur in a finite simulation involving only 100 particles? The only possibility is to prepare the system in a metastable state near the stability limit. In this situation the critical nucleus may be small enough that a destabilizing fluctuation would have a reasonable probability to occur (cf., the MC simulations of Ref. 52).

Careful studies of localized fluctuation effects using microscopic chemical simulations are only beginning at the present time. Nevertheless, some initial results have been obtained which are highly suggestive, although not yet definitive. With the same dynamical data and mean-field radius as before, several sets of RMD simulations were carried out in the neighborhood of the metastable states corresponding to $\eta = 5$.

The sequence of frames in Figure 6 shows a typical chemical evolution near the stability limit of the lower metastable branch. The behavior of an entire series of 11 runs at $\varepsilon = 2.348$ ($\varepsilon/\eta = 0.4696$) is displayed in Figure 8. Here the B-molecule population is plotted against collision number for the first 10^4 collisions. Within this relatively short time span a variety of types of evolution is found. All runs begin with 100 molecules in the form A. In this special case of a pure component fluid, both dynamical and chemical initial conditions are the same for the entire set. In addition the sequence of dynamical events (collisions) is the same, so that the different results are due to the sequence of random numbers which determine the chemical evolution. In more general situations, the runs begin with a chemical mixture of fixed composition. The initial identities of the molecules, assigned at random, are then different in each run, and this difference is reflected in the subsequent chemical evolution.

Of the 11 systems studied (in Figure 8), only 2 spent no time at all in the metastable state at $N_B \sim 16$. These systems developed critical nuclei in the early generation of B molecules, as indicated in "snapshots" of their evolution, and proceeded directly to the stable steady state at $N_B \sim 89$. The remaining systems all stabilized at the metastable state, 5 stayed there throughout the time shown in Figure 8, and one did not destabilize until after 5×10^4 collisions had occurred. The other 4 developed critical nuclei during the time of observation as indicated in Figure 8, and evolved to the stable state. The different transition rates reflect two factors, the different sizes of the initial supercritical nuclei, and the different effects of fluctuations, due to particle motion and reaction, on the growth of the nuclei. Due to the small numbers of molecules contained in a typical nucleus, these fluctuations tend to be large. In Figure 8 the fluctuations in N_B have been suppressed so the "average" evolution could be followed more easily. The range of excursions from the two steady states is indicated at the right border of the figure.

In Figure 6 the sequence of events in a typical nucleation run is shown in the form of snapshots of the two-dimensional mixture. Beginning as before with a pure-A "phase" (frame 1), the system evolves first to the metastable state (frame 2), where it remains for over 80,000 collisions. By collision 86,274 (frame 3), a supercritical nucleus has appeared. Thereafter this nucleus grows rapidly as the system evolves to the stable steady state (frame 4). This example was selected for display because it exhibits a particularly "long-lived" metastable state and, together with Figure 8, illustrates the wide range of induction times observed in the formation of (super-) critical nuclei.

A large number (hundreds) of RMD simulations have been performed near the limits of stability of the two metastable branches of Figure 7. Effects similar to those of Figures 6 and 8 were observed over a range of ε values in both cases. As would be expected, a general feature of these studies is an increase in critical cluster size, and hence also in induction time, as values of ε somewhat nearer the equal-areas line are chosen. Further details of the computer experiments, including quantitative features such as the distribution of induction times and the distribution in size and composition of "chemical clusters," together with an applicable nucleation theory of chemical instabilities, are subjects of active investigation at present, and will be reported elsewhere.

Higher-Level Simulations: Stochastic Chemical Kinetics

Microscopic simulations, such as the RMD method discussed here, are not suitable for applications involving large numbers of particles, long simulation times, or chemical systems with widely differing reaction rates for different processes. The latter restriction in particular implies that many chemical systems of laboratory interest, which are governed at the deterministic level by "stiff" differential equations (48, 49), cannot yet be studied by the RMD method, even for short times in small systems. Such situations can be handled effectively in a higher-level simulation, however, in which individual microscopic processes are still followed, but no longer at the molecular level. In such a method (20, 21) only the chemical evolution (in a homogeneous volume) is simulated directly. Exactly as in the Markovian stochastic description of chemical kinetics, it is assumed that details of the underlying molecular dynamics may be eliminated to good approximation, at least in locally homogeneous subsystems, by appropriate averages over the "microscopic stochastic variables" (cf., Ref. 17). Hence a reduced description is adopted in which the species populations are followed, either globally in a spatially homogeneous system, or in locally homogeneous subvolumes or cells if a global description is not justified. For a given system the relative rates of reaction and transport determine an upper limit on volumes which may be treated as macroscopically homogeneous. A lower limit is established by assumptions made in passing from the molecular to Markovian stochastic level of description (17, 20, 21).

Several methods for simulating the stochastic evolution of chemical systems have been employed in recent years (Ref. 21 and references therein). Of particular interest is a stochastic simulation algorithm developed from the Markovian stochastic formulation of chemical kinetics (20, 21). Within this framework the transition probabilities for various kinetic processes take the general form

$$W_{\mu}(t)\delta t = c_{\mu} h_{\mu} \delta t \tag{11}$$

Here h_{μ} is a combinatorial factor expressing the number of distinct ways that the species necessary for an R_{μ} reaction can be selected from the populations within homogeneous volume V at time t. The quantity $c_{\mu}\delta t$ is then the "average probability that a particular combination of R_{μ} reactant molecules will react in the next infinitesimal time increment δt" (21, 22). Usually c_{μ} and h_{μ} are simply related to the macroscopic rate constant k_{μ} and the product of reactant species populations, respectively, as one would intuitively expect from macroscopic considerations.

Turning now to the reaction probability function $P(\tau,\mu)$ defined earlier, we can write an exact expression based on the Markovian stochastic theory applied to chemical kinetics (20, 21)

$$P(\tau,\mu) = h_{\mu}c_{\mu} \exp\left(- \sum_{\nu=1}^{M} h_{\nu}c_{\nu}\tau \right) \tag{12}$$

Depending as it does on the instantaneous values of c_{μ} and h_{μ} for all M reactions, $P(\tau,\mu)$ is evidently a complicated function of the evolving chemical state of the system. Using this function, however, it is possible to construct a rigorous algorithm for simulating the time-evolution of chemical systems to which the fundamental stochastic hypothesis applies (21, 22). The algorithm involves a proper random (Monte Carlo) selection, at the time t of an evolution, of quantities τ and μ which completely specify the next chemical event which will occur. With normalized transition probabilities obtained from Eq. 11, the channel-selection proceeds as in the RMD simulations (cf, Eq. 9). The appropriate time interval τ is given in terms of a second uniform random number in the unit interval, β, by the relation

$$\tau = -W^{-1} \ln \beta \ , \quad W \equiv \sum_{\mu=1}^{M} h_{\mu}c_{\mu} \tag{13}$$

This selection process is then iterated, beginning from an initial state of the system, as defined by species populations, to simulate a chemical evolution. A statistical ensemble is generated by repeated simulation of the chemical evolution using different sequences of random numbers in the Monte Carlo selection process. Within limits imposed by computer time restrictions, ensemble population averages and relevant statistical information can be evaluated to any desired degree of accuracy. In particular, reliable values for the first several moments of the distribution can be obtained both inexpensively and efficiently via a computer algorithm which is incredibly easy to implement (21, 22), especially in comparison to now-standard techniques for solving the stiff ordinary differential equations (48, 49) which may arise in the deterministic description of chemical kinetics (53). Now consider briefly the essential features of a simple chemical model which illustrates well the attributes of stochastic chemical simulations.

Symmetry-Breaking Instabilities: The Trimolecular Reaction. The "Brusselator" or trimolecular reaction is the simplest model which exhibits instabilities that may be symmetry-breaking in space and/or time. Although it does not represent an actual chemical reaction, it is nevertheless the best-studied and most widely known theoretical model for chemical instability phenomena. Historically it is the model on which the study of dissipative structures was begun by members of the Brussels School of Thermodynamics (hence its popular name) a decade ago (44, 45, 46, 47). A detailed theoretical analysis of the Brusselator has been presented recently in a monograph by Nicolis and Prigogine (17). Relevant aspects of this development are reviewed in the following paragraphs:

Without loss of generality the four reaction steps of the model are taken to be irreversible, thereby establishing the system infinitely far from equilibrium:

$$\begin{aligned} A &\xrightarrow{k_1} X \\ B + X &\xrightarrow{k_2} Y + D \\ 2X + Y &\xrightarrow{k_3} 3X \\ X &\xrightarrow{k_4} E \end{aligned} \qquad (14)$$

With A and B assumed to be in large excess (or buffered) so that they act as time-independent reservoirs, the deterministic rate equations for this system become (for convenience the rate constants are set to unity)

$$\frac{\partial X}{\partial t} = A - BX + X^2Y - X + D_X \frac{\partial^2 X}{\partial r^2}$$
$$\frac{\partial Y}{\partial t} = BX - X^2Y + D_Y \frac{\partial^2 Y}{\partial r^2} \quad (15)$$

where r is the spatial coordinate for a one-dimensional system and D_X, D_Y are Fick-type diffusion coefficients for intermediates X and Y, respectively.

It is easy to see that this system admits a unique spatially homogeneous steady state solution ($X_0 = A$, $Y_0 = B/A$), which is the continuous extension of the thermodynamic branch into the far-from-equilibrium region. Using the familiar linear stability analysis, we find two types of situations in which this steady state may become unstable. If B is taken as the single independent variable (i.e., A, D_X, D_Y fixed), then the condition for an oscillatory instability is satisfied when B reaches

$$B' = 1 + A^2 + \frac{m^2\Pi^2}{\ell^2}(D_X + D_Y) \quad (16a)$$

and for a nonoscillatory instability when B reaches

$$B'' = (1 + \frac{\ell^2}{m^2\Pi^2}\frac{A^2}{D_Y})(1 + \frac{m^2\Pi^2}{\ell^2}D_X) \quad (16b)$$

where ℓ is the length of the system and the range of permissible values for the wave number m depends on the boundary conditions.

In order to understand better the nature of the instabilities possible at these two points, consider the "critical" wave number m_c for disturbances which will produce instability at the smallest value of B. Minimizing the expressions in Eqs. (16a), (16b) with respect to m yields

$$m_c' = 0 \qquad B_c' = 1 + A^2 \quad (17a)$$

$$m_c'' = \frac{\ell}{\Pi}\left(\frac{A^2}{D_X D_Y}\right)^{1/4} \qquad B_c'' = \left(1 + A\left(\frac{D_X}{D_Y}\right)^{1/2}\right)^2 \quad (17b)$$

In an actual chemical mixture there may be present spontaneous fluctuations having a wide range of wavelengths. As B increases, then, the instability which appears will correspond to the smaller of the two critical values B_c.

For an instability at B_c', Eqs. (17a) imply that very long wavelength perturbations will grow and oscillate, the system remaining spatially homogeneous. Before the instability point is reached, fluctuations either decay monotonically or perform damped oscillations about the steady state. Beyond instability, fluctuations are amplified, the system leaves the steady state, and evolves to an undamped periodic regime. The resulting dissipative structure is a unique (in this example) limit cycle which is asymptotically stable. That is, an initial point anywhere in the space X,Y tends to the same period trajectory (Figure 9). The important point is that the chemical reaction leads to coherent time behavior; it becomes a chemical clock.

It is clear from Eqs. 16 and 17 that the possible appearance of finite spatial inhomogeneities involves a "cooperation" between chemistry and transport. Indeed, whenever the rates of diffusion and chemical reactions are "comparable," short wavelength disturbances may grow. For an instability at B_c'', for example, the system amplifies a fluctuation of "critical" wave number m_c'' and ultimately reaches a stable nonuniform steady state (see Figure 10).

The behavior of this system beyond the instability at B_c'' has been the subject of extensive numerical study. In this way the first spatial dissipative structures were predicted using the Brusselator equations (17, 44, 45, 46, 47). In general, of course, the length of the system will not be an integral multiple of the critical wavelength obtained from Eqs. (17b). This leads to an indeterminacy in the nature of the spatial dissipative structure which bifurcates from the homogeneous steady state at B_c''. The system will select that wavelength, compatible with the boundary conditions, which lies nearest the wavelength of the particular disturbance which destabilizes the initial steady state. Once one or another dissipative structure is established, however, it is then stable with respect to small fluctuations, and hence in its wavelength possesses a primitive "memory" of the fluctuation from which it originated.

Other than the two simplest possibilities treated here, the Brusselator also admits more complicated situations, including time-independent, oscillatory, and travelling wave solutions of the governing macroscopic equations. When nonuniform distribution of the reservoir variable A is permitted, moreover, spatially localized structures of several types have been realized. The implications of all these predictions for self-organizing systems in chemistry, biology, ... are discussed in Ref. 17, as are the numerical (deterministic) simulations which verify the theoretical predictions.

Spontaneous Fluctuations at Chemical Instabilities via Stochastic Simulations. Let us now consider an application of stochastic simulation methods to the Trimolecular reaction. Here we consider a few realizations of the underlying stochastic process in order to illustrate the main ideas. The initial investigation focuses on the size-dependence of fluctuations near the homogeneous transition to limit cycle oscillations. The deterministic system is defined according to Eqs. 14-17, with A = 1 and $k_\mu = 1$

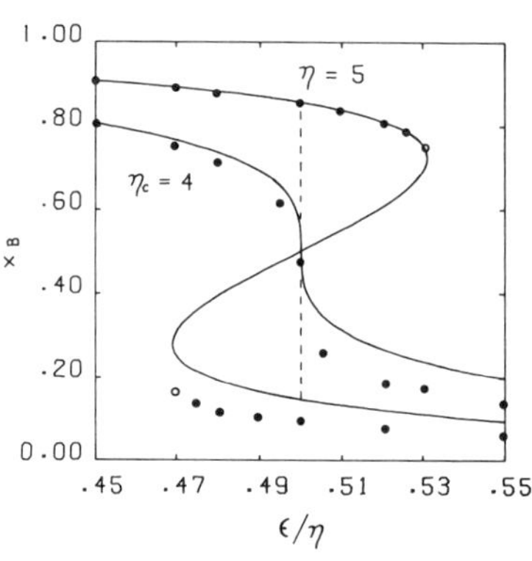

Journal of Physical Chemistry

Figure 7. Average composition (x_B) obtained from RMD simulations with $R_{mf} = 2.50\sigma$. Solid curves give the deterministic steady states, filled circles denote computer-experimental steady states which are stable during the time of observation, while open cricles denote metastable steady states of the computer model (22).

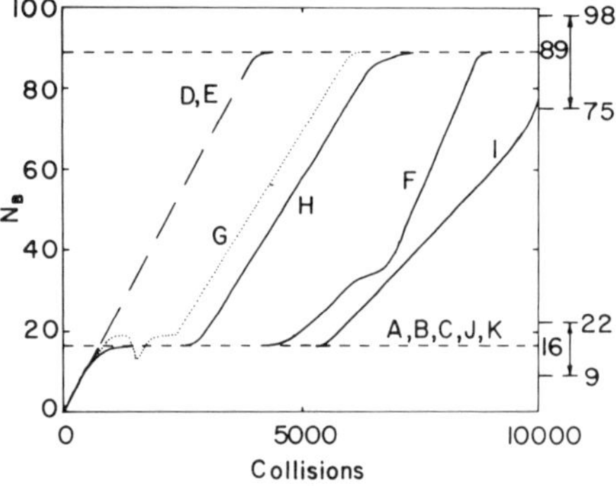

Journal of Physical Chemistry

Figure 8. Evolution of N_B in a set of 11 runs using common dynamical data. Same parameters and initial conditions as Figure 6. The first 10% of the full 10^5-collision simulation is shown. Fluctuations about the trajectories have been suppressed, but the range of fluctuations about the two steady states is indicated at right (22).

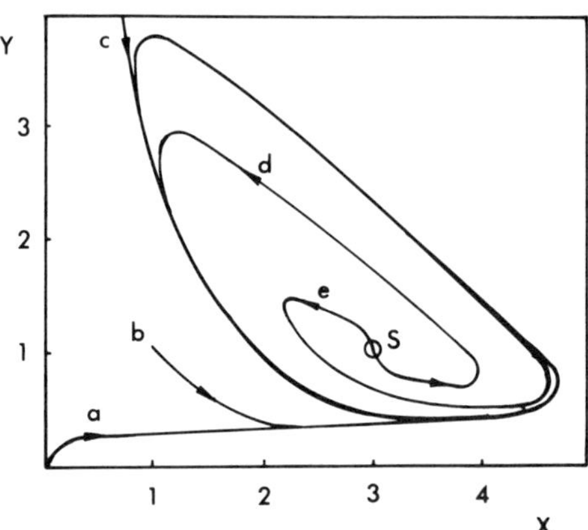

Figure 9. Approach to stable limit cycle from unstable steady state S and from other initial conditions (A = 1, B = 3, $k_\mu = 1$, $\mu = [1,4]$)

for μ = [1,4]. In a spatially homogeneous system, then, the transition to bulk oscillations will occur at the critical point given by Eq. 17a, or $B_c' = 2$. The object of the simulations is to study fluctuations in a stochastic version of this system, for values of B in the subcritical, critical, and supercritical regions, and to characterize the fluctuations as a function of the size of the system. If A denotes a number density (or molar concentration), then $N_A \equiv A \cdot V$ (or $A \cdot \Omega$), the number of molecules of type A is a direct measure of the system size. (The system size factor $\Omega = N_0 V$ for molar concentrations, where N_0 is Avogadro's number.) Hence the simulation strategy will be to pick a size N_A, which also determines the steady-state X population $N_X = N_A$, and vary B (i.e., N_B) through the region of interest.

The results of stochastic simulation for N_A = 500 are displayed in Figure 11. Here are plotted the populations N_X, N_Y against time in a, c, e, and the N_X - N_Y phase portrait in b, d, f, for three values of B. The steady-state values (N_X^0,N_Y^0) are indicated on each plot. The population data which appear in the figures represent time-dependent averages, taken over times long compared to the time between reactive events, but short on the time scale of any macroscopic evolution. Hence a high frequency (molecular) component has been averaged out of the raw population data.

In the far subcritical region (B = 1; Figures 11a,b) the system remains near the point (N_X^0,N_Y^0), but its random evolution tends to smear out the deterministic steady state. In the time trace fluctuations are seen to be small. As B increases toward B_c', the random motion about the steady state increases in amplitude until, at the critical point (B = 2; Figures 11c,d), occasional large amplitude quasi-oscillations appear in the otherwise chaotic motion about (N_X^0,N_Y^0). Beyond B_c', the oscillations gradually increase in coherence, but remain relatively erratic until the amplitude of oscillation greatly exceeds the background level of thermal fluctuations. In the far supercritical region (B = 3; Figures 11e,f), a regular pattern is evident, with large amplitude relative to fluctuations. Nevertheless, even here a degree of incoherence remains, a direct manifestation of the finite numbers effect common to all discrete simulations.

Following the series with N_A = 500, further simulations were performed with N_A values of 100, 1000, and 5000. The results of all the homogeneous simulations are collected in Figure 12. Here is plotted the average peak-to-peak amplitude of excursions in the X-population, normalized by the size parameter N_A to give an intensive quantity for purposes of comparison. This "reduced X-amplitude" is plotted as a function of B over a wide range which includes the critical point B_c' = 2. Each point represents an average over ten runs which differ only in the sequence of random numbers used in the channel-selection process. Error bars for several N_A = 100 points indicate the dispersion in the individual runs. As one would expect, this dispersion decreases with

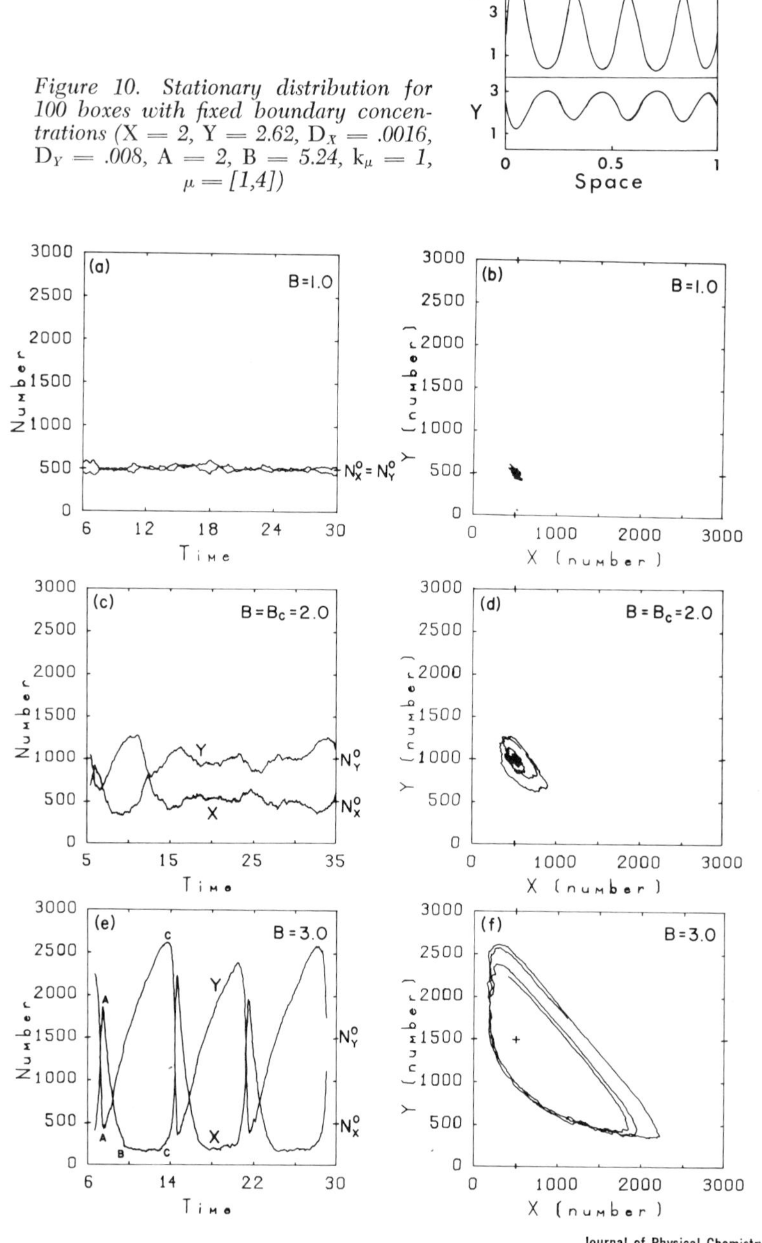

Figure 10. Stationary distribution for 100 boxes with fixed boundary concentrations (X = 2, Y = 2.62, D_X = .0016, D_Y = .008, A = 2, B = 5.24, k_μ = 1, μ = [1,4])

Figure 11. Time traces of the X, Y populations (a, c, e) and their associated phase portraits (b, d, f). The data are from stochastic simulation of the homogeneous trimolecular reaction, in the subcritical (a, b), critical (c, d), and supercritical (e, f) regions. A = 1, k_μ = 1 (μ = [1,4]), B'_c = 2; N_A = 500 (22).

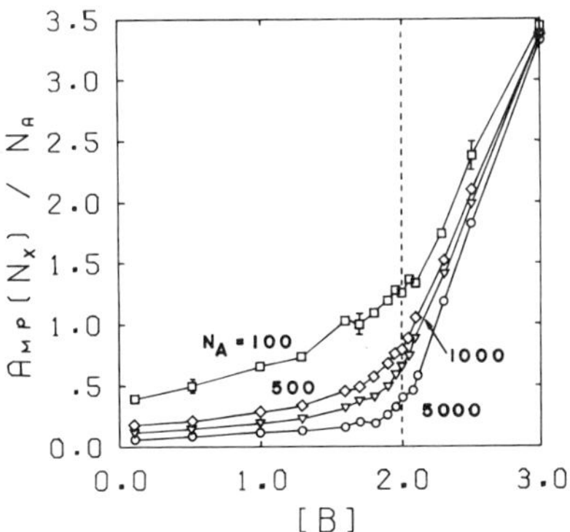

Journal of Physical Chemistry

Figure 12. Reduced N_X *amplitude near the second-order transition to homogeneous oscillations in the trimolecular reaction, as a function of the system size,* N_A*. The order parameter value at each experimental point is an average over 10 simulations, the error bars for* $N_A = 100$ *indicating the dispersion about the mean. Same macroscopic parameters as Figure 11 (22).*

increasing N_A.

Qualitatively, the second-order character of the transition to homogeneous oscillations is already evident in Figure 12. Below B_c', the finite values of the order parameter, as measured by $\Delta N_X / N_A$, reflect the effect of fluctuations about the homogeneous steady state. For $B \lesssim 1$, the fluctuations are found to obey the "square-root rule," to several decimal places accuracy, as would be expected far from a critical point. Similarly, for $B \gtrsim 2.5$, fluctuations about the macroscopic evolution behave in the "classical" manner. Near B_c', however, this simple size-dependence of fluctuations breaks down. Finally, the order-parameter data are seen to approach a definite limiting curve as the system size increases, indicating increased coherence and long-range temporal order. According to the phase transition analogy, we would expect to find perfect temporal coherence (i.e., a chemical clock) in the limit of a macroscopic homogeneous system.

Further stochastic simulation studies now in progress are concerned with fluctuation and nucleation in evolving chemical systems (e.g., limit cycle oscillations, combustion and explosions) and at the transition to spatial dissipative structure (cf., Figure 10). In the latter case, for example, stochastic simulations verify the existence of critical long-range spatial correlations predicted in a stochastic theoretical study of the model (cf., Ref. 17).

Concluding Remarks

As we have seen, the need for chemical simulations at a variety of levels is quite clearly demonstrated in the broad class of systems in which molecular and supermolecular interactions lead to macroscopic cooperativity, self-organization, and evolution. Although this class extends far beyond the domain of "classical" chemistry as well, it nevertheless covers the chemical spectrum, from combustion and explosions in gaseous mixtures to biochemical regulation of metabolism and biosynthesis, and from adsorption/chemisorption and heterogeneous catalysis to chemical oscillations, spatial structures, and travelling chemical waves in condensed phase chemistry and biology.

It is evident from the proliferating literature in this field (cf, Ref. 17) that application of macroscopic and statistical theoretical methods to chemical instabilities is far more difficult in general than for their equilibrium counterparts. In particular, the use of any analytical theory to study typical chemical systems of laboratory importance is out of the question, so it is precisely in this connection that discrete simulation methods may ultimately make their greatest contribution.

In addition, a number of important fundamental questions regarding nonequilibrium phase transitions are beyond resolution with available stochastic methods, and indeed require a molecular picture. Here again the answer is not readily available from theory. As one might expect from studies of analogous problems in equilibrium phase transitions and critical phenomena, nonequilibrium phase transitions present great difficulties in both microscopic theory and laboratory experiment.

In both experimental and theoretical studies, therefore, simulations involving actual molecular dynamics, or the higher level reactive molecular dynamics, may provide the necessary information at a not unreasonable expense. When absolute molecular detail is not required, or can be incorporated into average quantities to parameterize a higher-level simulation, then stochastic simulations provide a powerful yet flexible technique which should be applicable to a wide variety of problems. Because they are limited neither by number of variables nor by mechanistic complexity, such methods afford a practical means of treating highly complicated systems. Moreover, the problems of "stiffness" which plague deterministic methods do not arise in stochastic simulations. Consequently, for some problems, stochastic simulations may actually out-perform numerical methods based on macroscopic differential equations, particularly when large numbers of chemical species governed by stiff kinetic equations must be treated in two or three space dimensions. At present this conjecture must be qualified by the practical limitation of discrete event methods to relatively small systems for relatively short simulation times (e.g., up to 10^6 molecules for 10^6 to 10^8 events).

This paper has focused on two recent computer methods for discrete simulation of chemical kinetics. Beginning with the realization that truly microscopic computer experiments are not at all feasible, I have tried to motivate the development of a hierarchy of simulations in studies of a class of chemical problems which best illustrate the absolute necessity for simulation at levels above molecular dynamics. It is anticipated (optimistically!) that the parallel development of discrete event simulations at different levels of description may ultimately provide a practical interface between microscopic physics and macroscopic chemistry in complex physicochemical systems. With the addition to microscopic molecular dynamics of successively higher-level simulations intermediate between molecular dynamics at one extreme and differential equations at the other, it should be possible to examine explicitly the validity of assumptions invoked at each stage in passing from the molecular level to the stochastic description and finally to the macroscopic formulation of chemical reaction kinetics.

With key algorithmic elements drawn from familiar MD and MC methods and from a Markovian stochastic theory of microscopic rate processes, the two methods discussed here greatly idealize important microscopic features of the systems under study, and hence represent only the first few steps in the overall program. In this sense they may be reasonably viewed more as an extension "downward" from the macroscopic description than as a faithful representation of microscopic chemical dynamics. Appropriately, therefore, the initial applications have been to questions raised in deterministic (and stochastic) theories of chemical kinetics. As we have seen already, however, atomic and molecular detail can be incorporated into computer simulations by performing suitable averages over selected microscopic degrees of freedom. In this way effective force laws, for one example, or microscopically derived reactive transition probabilities (i.e., "rate constants"), for another, should permit simulations at a variety of levels of microscopic detail.

Abstract

For the broad class of systems in which molecular interactions lead to macroscopic collective behavior, it is often the case that neither microscopic theory nor laboratory experiment can provide unambiguous interpretation of observed phenomena. Faced with such an impasse, one may reduce the gap between theory and experiment by means of discrete simulations using high-speed computers. Such methods have long been invaluable in the study of classical fluids, for example, and in various forms are used in all disciplines which must treat complex interactions among large numbers of individual elements. Within chemistry the physical properties of solvents, surfaces, and complex molecules are already studied by means of computer "experiments," and further applications cover the chemical spectrum from combustion and explosions to polymerization and biochemical regulation, and from sorption and heterogeneous catalysis to chemical instabilities, self-organization, and evolution far from equilibrium. At the level of molecular dynamics, reactive collisions are rare dynamical events, so complete microscopic simulations of even simple chemical processes are not feasible. To the extent that the influence of nonreactive processes on chemistry can be either ignored or treated statistically, however, it is possible to develop higher-level simulations in which only events of direct chemical interest are followed. In this way a hierarchy of discrete simulations can span the gap between molecular dynamics and chemical kinetics, with additional computational efficiency gained at the expense of unwanted molecular detail, providing ultimately a practical interface between microscopic physics and macroscopic chemistry.

Acknowledgement

I wish to thank Professor I. Prigogine, Drs. J. Hiernaux and D. T. Gillespie, and Messrs. S. Kelkar and M. Malek-Mansour for stimulating discussions on the many facets of interfacing discrete simulations and chemical kinetics. I thank especially Dr. Hiernaux and Mr. Kelkar for their efforts in computational aspects of the RMD and stochastic simulations. This work was supported in part by the Robert A. Welch Foundation.

Literature Cited

1. B.J. Alder and T.E. Wainwright, "Proceedings of the International Symposium on Transport Processes in Statistical Mechanics" (Brussels, Aug. 27-31, 1956; I. Prigogine, Ed.) Interscience, New York, 1958, pp. 97-131.
2. B.J. Alder and T.E. Wainwright, J. Chem. Phys., 31, 459 (1959).
3. A. Rahman, Phys. Rev., 136, A405 (1964).
4. L. Verlet, Phys. Rev., 159, 98 (1967).
5. A. Rahman and F.H. Stillinger, J. Chem. Phys., 55, 3336 (1971).
6. D.L. Bunker, Methods Comp. Phys., 10, 287 (1971).
7. W.W. Wood, in "Fundamental Problems in Statistical Mechanics," Vol. 3 (E.G.D. Cohen, Ed.) North-Holland, Amsterdam, 1975, pp. 331-388.
8. J.R. Beeler, Jr., Adv. in Materials Research, 4, 295 (1970).
9. W.G. Hoover and W.T. Ashurst, in "Theoretical Chemistry--Advances and Perspectives," Vol. 1 (H. Eyring and D. Henderson, Eds.) Academic Press, New York, 1975, pp. 1-51.
10. W.W. Wood and J.J. Erpenbeck, Ann. Rev. Phys. Chem. 27, 319 (1976).
11. F.T. Wall, L.A. Hiller, Jr., and J. Mazur, J. Chem. Phys., 29, 225 (1958).
12. F.T. Wall, L.A. Hiller, Jr., and J. Mazur, J. Chem. Phys., 35, 1284 (1961).
13. D.L. Bunker, J. Chem. Phys., 37, 393 (1962).
14. J.C. Keck, Disc. Faraday Soc., 33, 173 (1962).
15. N.C. Blais and D.L. Bunker, J. Chem. Phys., 37, 2713 (1962).
16. N.C. Blais and D.L. Bunker, J. Chem. Phys., 39, 315 (1963).
17. G. Nicolis and I. Prigogine, "Self-Organization in Non-Equilibrium Systems," Wiley-Interscience, New York, 1977.
18. J. Portnow, Phys. Letters, 51A, 370 (1975).
19. P. Ortoleva and S. Yip, J. Chem. Phys., 65, 2045 (1976).
20. D.T. Gillespie, J. Comp. Phys., 22, 403 (1976).
21. D.T. Gillespie, J. Phys. Chem., 81, 2340 (1977).
22. J.S. Turner, J. Phys. Chem., 81, 2379 (1977).
23. N. Metropolis, A.W. Rosenbluth, M.N. Rosenbluth, A.H. Teller, and E. Teller, J. Chem. Phys., 21, 1087 (1953).
24. W.W. Wood, in "Physics of Simple Liquids" (H.N.V. Temperley, J.S. Rowlinson, and G.S. Rushbrooke, Eds.) North-Holland, Amsterdam, 1968, pp. 115-230.
25. G.A. Bird, Ann. Rev. Fluid Mech., 10, 11 (1978).
26. S. Glasstone, K.J. Laidler, and H. Eyring, "Theory of Rate Processes," McGraw-Hill, New York, 1941.
27. C.H. Bennett, in "Algorithms for Chemical Computation" (R.E. Christoffersen, Ed.) ACS Symposium Series, Vol. 46, American Chemical Society, Washington, D.C., 1977, pp. 63-97.
28. S.A. Adelman and J.D. Doll, J. Chem. Phys., 61, 4242 (1974).
29. S.A. Adelman and J.D. Doll, J. Chem. Phys., 62, 2518 (1975).
30. J.D. Doll, L.E. Myers, and S.A. Adelman, J. Chem. Phys., 63, 4908 (1975).

31. S.A. Adelman and B.J. Garrison, J. Chem. Phys., 65, 3751 (1976).
32. J.D. Doll and D.R. Dion, J. Chem. Phys., 65, 3762 (1976).
33. F.W. deWette, R.E. Allen, D.S. Hughes, and A. Rahman, Phys. Letters, 29A, 548 (1969).
34. W.C. Harrison and W.C. Schieve, J. Stat. Phys., 3 35 (1971).
35. H.W. Harrison, W.C. Schieve, and J.S. Turner, J. Chem. Phys., 56, 710 (1972).
36. H.W. Harrison, "A Molecular Dynamics Study of Dimer Formation," Ph.D. Dissertation, University of Texas, Austin, 1972.
37. H.W. Harrison and W.C. Schieve, J. Chem. Phys., 58, 3634 (1973).
38. W.C. Schieve and H.W. Harrison, J. Chem. Phys., 61, 700 (1974).
39. M. Synek, W.C. Schieve, and H.W. Harrison, J. Chem. Phys., 67, 2916 (1977).
40. W.H. Zurek and W.C. Schieve, Comp. Phys. Commun., 13, 75 (1977).
41. W.H. Zurek and W.C. Schieve, J. Chem. Phys., 68 (3), 840 (1978).
42. W.H. Zurek and W.C. Schieve, Phys. Letters, A (in press, 1978).
43. J.O. Hirschfelder, C.F. Curtiss, and R.B. Bird, "Molecular Theory of Gases and Liquids," John Wiley, New York, 1954.
44. I. Prigogine and G. Nicolis, J. Chem. Phys., 46, 3542 (1967).
45. R. Lefever, G. Nicolis, and I. Prigogine, J. Chem. Phys., 47, 1045 (1967).
46. I. Prigogine and R. Lefever, J. Chem. Phys., 48, 1695 (1968).
47. R. Lefever, J. Chem. Phys., 49, 4977 (1968).
48. C.W. Gear, Commun. Am. Comput. Mach., 14, 185 (1971).
49. D.D. Warner, J. Phys. Chem., 81, 2329 (1977).
50. T.L. Hill, "Introduction to Statistical Thermodynamics," Addison-Wesley, Reading, Mass., 1960.
51. I. Prigogine, R. Lefever, J.S. Turner, and J.W. Turner, Phys. Letters, 51A, 317 (1975).
52. A.I. Michaels, G.M. Pound, and F.F. Abraham, J. Appl. Phys., 45, 9 (1974).
53. Proceedings of a "Symposium on Reaction Mechanisms, Models, and Computers," 173rd Nat. Mtg. of the Amer. Chem. Soc., March 20-25, 1977, published in J. Phys. Chem., 81, no. 25 (entire issue)(1977).
54. M. Rao, B.J. Berne, and M.H. Kalos, J. Chem. Phys., 68 (4), 1325 (1978).

RECEIVED September 7, 1978.

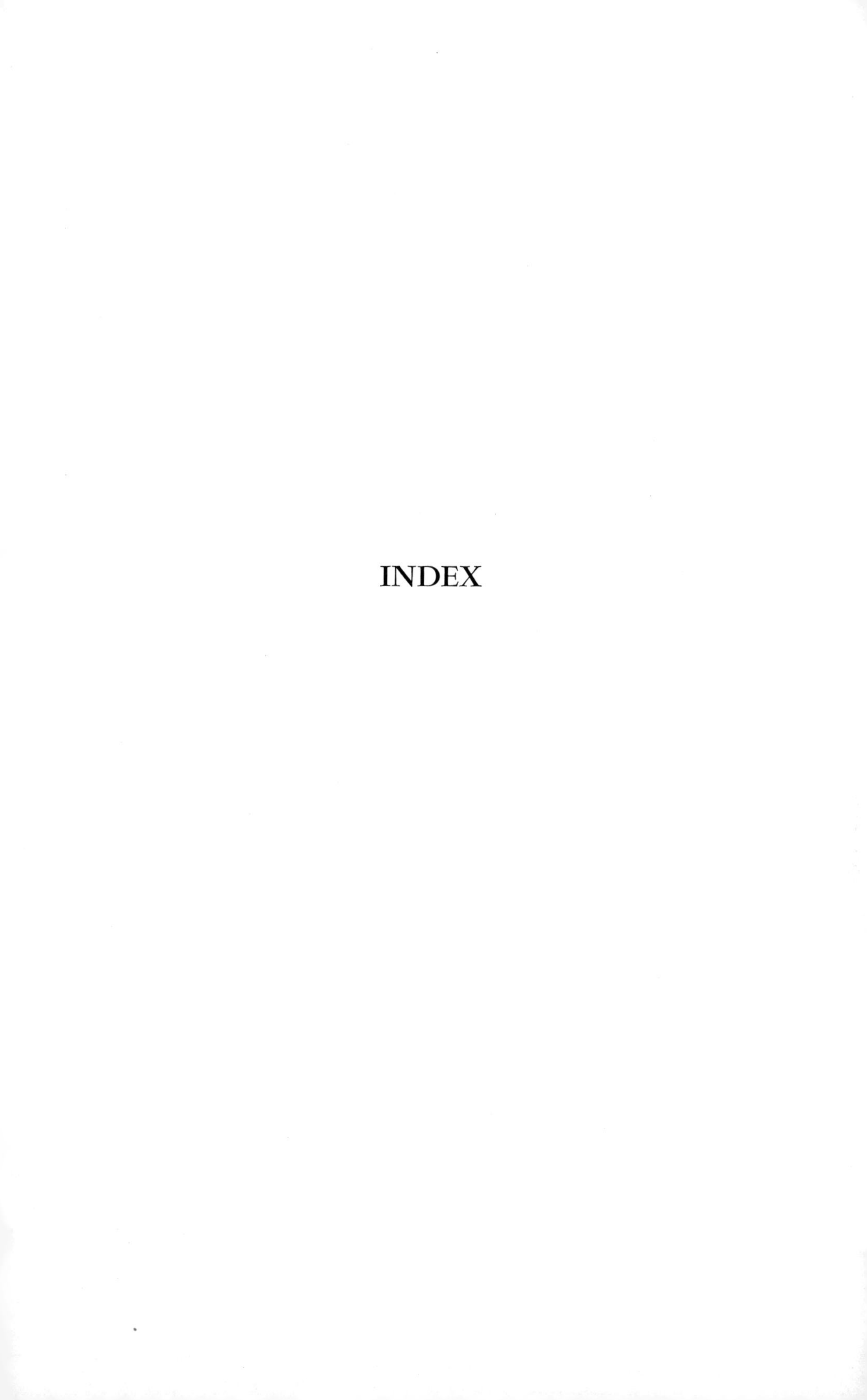

INDEX

INDEX